Designing Engineers

Inside Technology
edited by Wiebe E. Bijker, W. Bernard Carlson, and Trevor Pinch

Designing Engineers

Louis L. Bucciarelli

The MIT Press
Cambridge, Massachusetts
London, England

First MIT Press paperback edition, 1996

© 1994 Massachusetts Institute of Technology

This book was set in Baskerville by DEKR Corporation, Woburn, Massachusetts.

Library of Congress Cataloging-in-Publication Data

Bucciarelli, Louis L.
  Designing engineers / Louis L. Bucciarelli.
    p. cm—(Inside technology)
  Includes bibliographical references (p.).
  ISBN 978-0-262-02377-1 (hc : alk. paper), 978-0-262-52212-0 (pb) 1. Engineering design. I. Title. II. Series.
TA174.B756 1994
620′.0042—dc20   94-17348
                                                               CIP

The MIT Press is pleased to keep this title available in print by manufacturing single copies, on demand, via digital printing technology.

Contents

For Ann

Preface

This book is a study of engineers at work. It is based, in part, on my experiences consulting and working in a variety of settings. Except for the time I was asked to review the redesign of a nineteenth-century waterwheel at Old Sturbridge Village, almost all of my associations have been with high-tech firms and laboratories. In the past ten years I have tried to infuse my engineering assignments with a broader purpose—namely, to understand and explain the engineering design process in more than the usual instrumental terms, which struck me as seriously deficient as a basis for understanding how engineers actually work.

Some of what appears here is drawn from earlier essays: a 1988 piece entitled "An Ethnographic Perspective on Engineering Design" (*Design Studies*, July 1988, pp. 159–168), a conference paper on "Engineering Design Thinking" (*Proceedings of the 1987 ASEE Annual Conference*, Reno, Nevada), and a chapter in a book edited by Frank Dubinskas (*Making Time: Culture, Time, and Organization in High Technology*, Temple University Press, 1988). I finished a first draft while on a sabbatical leave during the academic year 1988–89. Bruno Latour and his colleagues at the Ecole des Mines, Paris, were most congenial hosts for this period, and I am also indebted to Terry Shinn of CNRS for his willingness to critique my writing in its earliest stages and for his arranging support from CNRS in the summer of 1990 so that we could begin a comparative study of engineers at work in France and the United States.

Support for the early stages of this work was provided by the National Science Foundation through its Ethics and Values in Science and Technology program. That support gave some needed legitimacy to the project and brought me in contact with Sharon Traweek, who became my tutor in anthropological methods and perspectives. Cross-disciplinary exchanges of substance and significance are unfortunately rare events at the university, and when they occur as part of a research project, they must be valued even more highly. Professor Traweek guided me through the literature and taught me the nature of ethnographic data, in all of its forms.

Another source and inspiration for this book was a joint project with Don Schön to study and compare design processes in architecture and in engineering. That work, too, was funded by NSF; and it led to several conference papers, extracts from which appear in these pages. Professor Schön's critique and encouragement of my attempts to formulate a clear and substantial explanation of what I meant by the phrase *object world* were key to my progress.

Colleagues in MIT's Program in Science, Technology and Society gave some of my first chapters a thorough reading. Sherry Turkle, Lisa Rofel, and more recently Michael Fisher did not hesitate to let me know when my writing went "over the top" or my analysis needed further attention.

Phyllis Klein provided much-needed assistance in the preparation of the manuscript. Trevor Pinch, my contact editor for the Inside Technology series, took responsibility for guiding me from first draft to the one you see here, and Larry Cohen at MIT Press made this journey relatively painless, indeed an occasion for learning in itself. I thank them for their attention and support.

Finally, I thank those participants in the design efforts I studied for allowing me to question and observe as they went about their day-to-day activities.

1

Introduction

This is a book about engineering design, but it is not a picture book. I will offer no illustrations of automobiles sporting sleek aerodynamic shapes, or of glistening kitchen appliances, or of sky-view offices with dispersed computer monitors and attractive people lounging about. Such visions rapidly become dated, like fashions in clothes or toothpaste tubes. I will take my chances with words; they endure somewhat longer.

This is a storybook in a way—a story about three design projects, two of which I observed firsthand. Like an ethnographer invited not just to dinner but to help out with the shopping, chopping, and peeling, I was able to participate in the design process at an engineering firm that produced photovoltaic modules and another that made x-ray inspection machinery for airports. My third story, based on an extended consulting experience, describes the design of a subsystem of a photoprint processing machine, the kind you see in storefronts, often bearing a foreign label, turning out four-by-six-inch enlargements of Uncle Charlie in front of the Eiffel Tower.[1]

I have included some diagrams and sketches—the sorts of visual representations participants in design make and use on their way to the production of automobiles, kitchen appliances, or computers. Although these visual images are important parts of the process, I want to emphasize that the type of designing I am interested in goes well beyond making automobiles look sleek, computer terminals efficient, or nuclear reactors unobtrusive. Though appearances can be central to the people specifically responsible for a product's looks, most participants in the design process worry primarily about how things "perform" or "behave." Looks are secondary; function is primary. Designing in this sense is about how things work.

Moreover, in the contemporary engineering firm, designing engages more than a lone engineer at a drafting board or work station. The design of a photovoltaic module, or an x-ray inspection system, or a new paper handler for a photoprint processor engages a wide variety of people within the firm: research scientist, marketing chief, lab technician, systems engineer, project manager, production engineer, purchasing agent, inventory controller. All can and do influence the design, and all must come to agreement in order to realize the design. The process is thus social, the business of a subculture. Not surprisingly, participants' visions of the social process of designing are strongly influenced by their understanding of the way the things they are designing work. To participants in design, the object serves as a kind of icon that embodies a set of attitudes and ways of thinking that are peculiar to engineering.

This book will thus focus on both object and process: object as embodiment of the way participants understand how things work, process as the way they go about designing. The two overlap, of course. Within the three design processes I observed, it was often difficult to draw a sharp line demarcating where "hard" object ended and "soft" social process began.

I want to start with the basic question of how objects work. This is more problematic than it might appear at first blush. Consider, for example, the telephone.

Do You Know How Your Telephone Works?

A few years ago, I attended a national conference on technological literacy, a topic that had recently attracted attention in the media as well as among scholars concerned with education at both secondary and postsecondary levels in the United States. One of the main speakers, a sociologist, presented data he had gathered in the form of responses to a questionnaire. After detailed statistical analysis, he had concluded that we are a nation of technological illiterates. As an example, he noted how few of us (less than 20 percent) know how our telephone works.

This statement brought me up short. I found my mind drifting and filling with anxiety. Did I know how my telephone works?

I squirmed in my seat, doodled some, then asked myself, What does it mean to know how a telephone works? Does it mean knowing how to dial a local or long-distance number? Certainly I knew that much, but this did not seem to be the issue here. Might it mean knowing how to install a phone, or perhaps knowing what to do when something goes wrong—no dial tone, noise on the line? But most of us do know that in such cases we need only call a service department. No, I suspected the question was to be understood at another level, as probing the respondent's knowledge of what we might call the "physics of the device."

I called to mind an image of a diaphragm, excited by the pressure variations of speaking, vibrating and driving a coil back and forth within a magnetic field, generating a variation in voltage across the terminals of the coil. I could even picture this signal traveling along the wire coming out of the phone into the wall and out onto the pole outside my house. But what then? How do all those different signals from different phones get sorted and routed to their proper destinations? I had no explanation for this.

If this was what the speaker meant, then, he was right: Most of us don't know how our telephone works, and I bet if he asked, most of us don't give a damn about knowing how it works as long as it does. Despair and guilt had yielded to cynicism.

Indeed, I wondered, does *he* know how *his* telephone works? Does he know about the heuristics used to achieve optimum routing for long-distance calls? Does he know about the intricacies of the algorithms used for echo and noise suppression? Does he know how a signal is transmitted to and retrieved from a satellite in orbit? Does he know how AT&T, MCI, and the local phone companies are able to use the same network simultaneously? Does he know how many operators are needed to keep the system working, or what those repair people actually do when they climb a telephone pole? Does he know about corporate financing, capital investment strategies, or the role of regulation in the functioning of this expansive and sophisticated communication system?[2]

Does *anyone* know how their telephone works?

How about the designer of the device? Surely, he or she must know. I followed this line of thought for a while, but again, I found my footing less sure than I had anticipated. Indeed, from my own observations, I can claim fairly confidently that there is no single individual alone who knows how all the ingredients that constitute a telephone system work together to keep each of our phones functioning. There is no one "maker." Instead, inside each firm, there are different interests, perspectives, and responsibilities—corporate planning, engineering, research, production, marketing, servicing, managing—and consequently different ways in which the telephone "works."

At this point I retreated from cynicism. The question now struck me as interesting, not as an instrument for testing technological literacy, but in its own right: What does it mean when someone claims to know how their telephone works? I conjectured that there could well be no unique criterion for judging responses; there could be as many legitimate, that is to say accurate, ways to describe how the telephone works as there are respondents.[3]

The narrow view of the workings of the telephone has the quality of a myth. But while the story about vibrating diaphragms and coils moving in a magnetic field may provide coherence amid complexity, may give us the confidence to respond in the affirmative to a sociologist on the track of technological literacy, and may encapsulate some particular facet of technical truth, taking it as *the* measure of a person's understanding of how today's sophisticated and dynamic system of communication works is naive.

This naivete is just one failing of the sociologist's research program; another is the presumption that you can test a person's "technological literacy" by literary means alone. Thus, if a respondent has just finished reading about the telephone in the *Encyclopedia of Science*, the sociologist would no doubt find him or her technologically literate. But this is neither sufficient nor necessary as a display of knowing how the telephone works. Indeed, it is little more than a test of reading retention.

No, the "knowing how it works" that has meaning and significance is knowing how to do something with the telephone—how to act on it and react to it, how to engage and appropriate the technology according to one's needs and responsibilities. Thus the everyday user around the house from the age of seven to seventy, the person who repairs the lines, the new product engineer, and the corporate executive each appropriates the telephone to his or her own interests, and each knows how it works.

Refocusing in this way, dragging myself away from an instrumental story derived from a vision of the scientific origins of the device, led me to reformulate the question. Rather than "Do you know how your telephone works?" I would ask, "How does the telephone symbolize, alienate, serve, or have meaning for you?" With this question we might be able to test the breadth and depth of a person's technological understanding and competence. Note, too, how, with this reorientation, the object loses its grip and the social context of its workings assumes primacy.

It is the fixation on the physics of a device that promotes the object as an icon in the design process. For while different participants in design have different interests, different responsibilities, and different technical specialties, it is the object as they see and work with it that patterns their thought and practice, not just when they must engage the physics of the device but throughout the entire design process, permeating all exchange and discourse within the subculture of the firm. This way of thinking is so prevalent within contemporary design that I have given it a label—"object-world" thinking—and will now detour a bit to give a preview of what I mean by the term. I will start with a less complex and more ancient technology and explore what it might mean to know how it works from the perspective of an object world.

How a Chair Works—One Perspective

Clearly we all know how to sit in a chair, so in this respect we know how it works: A chair is for sitting. That is how it functions. Unfortunately, this description is akin to saying that a telephone is for making telephone calls. Surely we can do better. (You never know when a sociologist is going to telephone to ask if you know how your chair works.) How about this?

A chair is a supported seat. A seat is a two-dimensional, finite, contoured surface, made of a rigid or flexible material, which provides immediate support to at least the buttocks of the human body. This support is sufficient to enable a person, when seated, to lift both feet off the ground without falling over. The surface is supported by a structure intervening with the ground.

A skeptic might interject here, "Does this mean that a hammock is a chair? Or a bed? Or a horse?" In response, we can start adding details. We might, for example, say that a chair is made of inanimate material. There goes the horse.[4] Or we can pass the buck back to a simpler precedent and build upon that:

A chair is a stool with a backrest, and a stool is a board elevated from the ground by supports.[5]

But this is still not good enough, for we still have to explain how a stool works. Why do most stools have three legs while a chair has four? We must strive to be even more scientific:

A chair is a four-legged stool with a back. Although three legs, or points of support, are sufficient to support the seat, the stability of seating is enhanced if we add a fourth point of support since, for said stability, a line connecting the center of gravity of an extended

body (yours or mine) to the center of gravity of the earth (or whatever planetary body one happens to be seated upon) must intersect the plane through the feet of the legs of the stool or chair, at
a point interior to the three- or four-sided plane figure having the
feet as vertices. With four legs, one can assure stability with a
square seat.[6]

This, then, is a chair and how it works. We can call this the principle
of operation of the chair—what Robert Pirsig would call its *underlying
form*.[7] This is the "physics of the device" knowledge that is often taken
as the hallmark of technological literacy. It constitutes a sparse, efficient,
generic, abstract identity of a chair, just as the identity of a circle is
defined by the Platonic ideal of "all points equidistant from a fixed
point."[8]

Given this principle, we can imagine an infinite variety of particular
embodiments that would function as a chair. A chair can be finely carved
or sparse yet stately. We can add arms and cushions to create a throne,
or set the thing up on curved rails to make a rocker—a matter of
controlled stability—if we want to get frivolous.

I don't claim that this generic description is the best possible in any
sense. Nor is it unique. Indeed, there are other object worlds we should
enter to define more fully how a chair works. We can talk about the
forces borne by the legs, which distort and deform relative to the seat
when a person sits down. We can even do a finite-element analysis on
the computer and estimate the internal stresses in the legs, the seat, and
the back. And we certainly ought to describe the craft knowledge of
chair object worlds: how to join the legs and back to the seat, whether
or not it is preferable to make the back an extension of the rear two
legs. All of this is part of an object world perspective on how a chair
works.

Historical Roots—Object Worlds

But this is neither a physics text nor a chair design manual. My intent
instead is to illustrate a particular way of thinking about and framing
an answer to the question "How does it work?" This way of thinking will
prove essential to understanding the thought and practice of participants in contemporary engineering design.

This has not always been the case. The first chair makers did not need
to enter the world of geometry, consider the conditions that ensure

static equilibrium, or call upon mathematical theories describing stress and strain within beams, legs, and arms in order to do their work. Vernacular technique was all they needed. One might claim that the underlying form of the chair was embedded in the crafter's rules of thumb, sense of symmetry, feel for an awl in making a cut, or know-how about joining wood to wood, but that knowledge remained unarticulated, tacit, and sensual (as far as we know). It was not of a kind with contemporary scientific understanding—for example, a Ptolemaic astronomer's explanation of planetary orbits, which relied upon the abstract notion of uniform circular motion and the machinery of epicycles and deferents (the elements of underlying form of the motion of the planets) and was drawn out in texts for all to read who could. But then, until relatively recently, the heavens were of a different nature than the stuff of the sublunary region. Indeed this "underlying form" way of thinking about earthly, mundane things like chairs and telephones was very much a child of the Renaissance. With the shattering of the perfection of the heavens, with the connection of the terrestrial to the celestial (the earth is not below, on center; the sun has spots; the planets are not wandering stars and themselves have moons; our moon is mountainous and, like Newton's apple, is falling toward the earth), the decaying stuff of this world became a candidate for abstract philosophical thought and discourse. Some attribute to Galileo this new way of framing and responding to the question.

Figure 1 shows Galileo's cantilever beam. He wanted to know how it works in object-world terms. That is, he was seeking, and felt he had found, its underlying form. He made the following claim:

A solid prism or cylinder of glass, steel, wood, or other material capable of fracture, which suspended lengthwise will sustain a very heavy weight attached to it, will sometimes be broken across . . . by a very much smaller weight, according as its length exceeds its thickness.[9]

Already we note the generic nature of his vision of "glass, steel, wood, or other material capable of fracture" and the instrumental relationship, somewhat hidden to the modern eye, in the proposition "be broken across . . . by a very much smaller weight according as its length exceeds its thickness." His demonstration immediately follows:

Let us imagine the solid prism ABCD fixed into a wall at the part AB; and at the other end is understood to be the force of the weight E. (Assuming always that the wall is vertical and the prism or cylinder is fixed into the wall at right

Figure 1

angles.) It is evident that if it must break, it will break at the place B, where the niche in the wall serves as support, BC being the arm of the lever on which the force is applied. The thickness BA of the solid is the other arm of this lever, wherein resides the resistance, . . . the moment of the force applied at C has, to the moment of the resistance which exists in the thickness of the prism (that is, in the attachment of the base BA with its contiguous part), the same ratio that the length CB has to one-half of BA. Hence the absolute resistance to fracture in the prism BD (being that which it makes against being pulled apart lengthwise) . . . has, to the resistance against breakage by means of the lever BC, the same ratio as that of the length BC to one-half of AB. . . . And let this be our first proposition.

The figure is striking—a fine etching in the Renaissance genre of nature observed. But it is also disconcerting. There is a lack of proportion in the weight hooked to the end of the beam relative to the beam itself. And the wall looks like the decaying remnant of some edifice that was perhaps once grand. How is it that the structure remains intact, supporting this massive weight, while the wall at the root of the beam seems fallen to such a sorry state? And the letters of the alphabet—what are they doing in this landscape?

The image is not to be read in this naturalistic way. No, it is a new kind of figure, a new image to go with a new science. Galileo means to help the reader follow his argument—that the beam will fail at the end B if the weight at the end C becomes too large. But there is more to it than that. The reader must see in the figure an angular lever with its fulcrum at B, one arm extending out along the beam to the weight at the end C, the other, shorter arm, extending from B up to A. Then the reader must imagine an internal force, a "resistance" acting over the arm AB, directed to the left, and visualize how this resistance would balance the weight suspended at E through the action of the lever. From the principle of equilibrium of the lever one obtains that the ratio of the force that will break the beam when applied transversely at E to the force that will break the beam when applied longitudinally at AB—a force Galileo calls "the prism's absolute resistance to fracture"—is the same as the ratio of one-half the distance AB to the distance BC.

This is how we are meant to read Galileo's proposition. He is explaining how a cantilever beam works by revealing its underlying form. His explanation is not a proof, though. The law of the lever was already well known. What is significant and innovative is the association of the workings of a load-bearing beam with the geometry and principle of the lever. The new science comes in positing a relation between the beam as physical artifact and vernacular technique and the beam as abstract lever and mathematical artifice.

As a *physical artifact,* the beam has a "thickness," a "length," a "mortise," and a "weight" at the end. These are significant, but the embellishments of these characteristics—the crumbling nature of the wall, the ivy, the grain in the wood, even the material from which the beam is made—are not.

As an *abstract lever,* there is a fulcrum at B, a force applied at C, and an (internal) resistance that "opposes the separation of the part BD lying outside the wall from that portion lying inside," and they bear the same ratio as half the length (lever arm) BA to the length (lever arm) BC. That is how the beam works.

Indeed, that *is* a cantilever beam. With Galileo, the questions "What is it?" and "How does it work?" collapse into one. When put, they solicit the same response. The attempt to answer either brings out the fact that a cantilever beam is a lever, just as a chair can be seen as a stable platform. In this, "we take for true being that which is only a method."[10]

Who Defines the Workings?

Seeking underlying form is but one approach to defining how a telephone, a chair, or a cantilever beam works. It guides participants designing within object worlds but doesn't always make sense for those of us who don't engage technology in this intimate way. Indeed, it is not always relevant, or all there is to the workings of technique; and sometimes it sounds downright silly to speak of a chair in this way.

For example, as you stand before Napoleon's throne in the Louvre, you are probably not thinking about underlying form. Or if you encounter a Rietveld chair in the Centre Pompidou, with its bright red back, glossy black seat, and its spindly arms, legs, and interlacing struts all tipped with yellow, you might think about underlying form only if it crosses your mind to wonder whether or not it would collapse if you sat on it. Or if you spot a Le Corbusier lounge chair in the window of a fashionable shop along the Rue Rivoli, priced at 12,380 francs, its inner constitution is not likely to be the first thing that passes through your mind. You don't spend $2,000 on a four-legged stool with a back.

A Rietveld or Corbusier chair or Napoleon's throne "works" as something other than what you or I am seated on at the moment. They work as art, as fashion, as symbol of power and prestige. Their primary function is not to support bodies. Their one common characteristic is that they are uncommon. Each has its own special identity that renders irrelevant our generic and abstract description. Our object-world definition of chair is meaningful only if we are talking about an *anonymous* chair. The anonymity of this chair is of a piece with the generic and abstract response we give to the question of how it works. It is no coincidence that Galileo's vision takes root at the first glimmerings of an industrial revolution.[11]

Far removed from the Renaissance, most of us don't see technology in terms of underlying form. We live in a world where each object appears in its particularity, as an artifact made of hardware or as a system of rules and constraints. The underlying form remains hidden, sometimes intentionally; our perceptions of how a telephone, a chair, a photovoltaic module, or an x-ray inspection machine function can be correspondingly "superficial." Yet we do know how they work, as shown by our knowing how to pass them by, work through them, remake them, or use them, even if but briefly.

Not all relations are superficial. Often the connection is dictated. A worker in a machine shop who tends a semiautomatic tool, which requires the insertion, then removal of the workpiece in but a second or two, has little choice in how to relate to technique. Other relationships are direct and intense by choice. Bricolage and do-it-yourself are possible, as Pirsig has so beautifully demonstrated in *Zen and the Art of Motorcycle Maintenance.* For example, an owner's careful and constant ministrations might cause a motorcycle or automobile to reveal a personality; or an office worker might become enthused about "hacking," learning to navigate the computer system in a personal and deliberate way. Moreover, even an ephemeral connection can be quite intense, as when a family reacts to the siting of a recycling facility across the town field; they may never have seen one, but they still have enough of a sense of how one might work to want to know more before they go along with the idea.

The way in which one sees how technology works is very much a matter of the nature of the encounter—whether it is in passing, intense in bricolage or dictation, or lay-political. Our relations to and hence our perspectives on technology may vary, but in general, as user, traveler, player, viewer, or tender, we do not have the same connection to technology that its makers have. However we appropriate technique, most of us do not see technology in terms of its formal structure, underlying form, or *inner constitution.*[12]

Those who are concerned with technological literacy often seem to be claiming that this is a sad state of affairs. Others see a deeper problem in our relations with technology, one that won't necessarily be solved by a universal program of education. They claim that we have embraced a new brand of slavery as we accept subservience to technique over which we have little control, much less understanding. Or, to put this idea in terms of a less odious metaphor, technology now legislates and structures our lives, yet we seem not to have been given the opportunity to vote.[13]

How does your telephone work? How does your chair really function? These questions fade in significance next to those that ask about legislating and ruling lives. Even if we can answer the sociologist "correctly" and thus be counted among the technologically literate, we remain uncomfortable: What disturbs us is the thought that, whatever our level of technological literacy, we are powerless to affect the form and function of the machinery that surrounds us, aids and comforts us, alienates and confuses us, improves our productivity, makes too much noise, sometimes leaks, can be beautiful, can be ugly, is often indispensable. . . . There it is: the telephone rings, and we answer.

It is a small step from these thoughts to a grand vision of technology that is autonomous in its workings, running our lives and out of the control of not just us ordinary citizens but engineers, managers, and corporate strategists who themselves don't fully understand what they are doing.

Two sorts of questions can be raised at this point. (1) Does the object dictate the nature of the relations we have with technique? Are the ways in which we relate to a technology, not just the ways by which we perceive it but a good bit of our life's fabric, inherent in the technology and its workings? (2) What and who determines the form and function of the machinery, structures, and systems that are so much a part of contemporary life? Who fixes the workings?

It is the second type of question that I want to address, for if technology is indeed a legislator in our lives, then we would like to put ourselves in a position to decide how it is to legislate. If alternative designs are possible, we would like to know as much and be let in on the choosing.

The form of the question as it stands is problematic, for if the designs of technology are determined uniquely by the way in which participants in the design process see them working—that is, according to underlying form and scientific principle—then who are we to argue with, or contribute to, that process? I will therefore rephrase the question more specifically: Does this instrumental, object-world way of thinking define the artifact, fix its form, and dictate all the ways in which it might work?

I have already claimed that this way of thinking strongly influences the process of design in that it pervades the day-to-day efforts of participants in that process. If we left it there, my answer would have to be yes, instrumental reasoning does fully define the product of design by framing all that transpires in the process of designing. But we can't leave it there. We must in fact distinguish between the way participants in design think the process of designing works and the way it actually does work.

Discrepancies like this are a major concern to those sociologists and anthropologists who seek underlying form in the everyday practices of a culture:

If the structure can be seen, it will not be at the earlier, empirical level, but at a deeper one, previously neglected; that of those unconscious categories which we may hope to reach, by bringing together domains which, at first sight, appear disconnected to the observer: on the one hand, the social system as it actually works, and on the other, the manner in which, through their myths, their rituals, and their religious representations, men try to hide or to justify the discrepancies between their society and the ideal image of it which they harbor.[14]

My goal in this book is to probe the function of the discrepancies between the social process of design as I have observed it, full of uncertainty and ambiguity, and the ideal image of an instrumental process according to which participants claim to work.

Visions of Process

Others have looked at the design process, or at invention and innovation, or at the diffusion of technology, and sought to explain how and why technology takes the form we see and experience. There are two basic perspectives on the topic, and I like to characterize them as the perspectives of the *savant* and the *utilitarian.*

For the savant, scientific principle and underlying form are indeed determinate, in the sense that the prevailing paradigms of the basic sciences—physics, chemistry, and now biology—are seen as the source and essence of technique. Scientific discovery becomes the determinant of technology's function and form; technology becomes a consequence of science. Furthermore, the process by which an innovative idea is translated into an artifact or system is itself seen as guided by the norms and laws of science. Good engineering practice and synthesis in design are viewed instrumentally, as applied science, with an allowance made for irrational behavior in the form of a healthy dash of creativity at the outset of the process.

In the mind of the savant, the actions, interests, and needs of corporate managers, consumers, stockholders, and workers—including the participants in design themselves—matter little. They lie somewhere outside the picture, although those that take this point of view may refer to "benefits for all humankind" and, when unsure of what those benefits might be, like to quote Ben Franklin—"What good is a newborn baby?" From this perspective, the call for technological literacy might be interpreted as expressing a need for more scientists and scientifically educated entrepreneurs who have the potential to invent things that will ultimately improve our well-being.

A savant who tries to reconstruct the design process of a device invariably focuses on the completed, fully functioning artifact. He or she will, for example, inspect a finished x-ray machine, analyze its internal logic, and then isolate the sequence of design decisions that explain how it came to be the way it is. The story will sound rational, but it will usually have little to do with the real-world process by which the artifact was given its form.

Standing before the machine, its deterministic functioning so dominates our thought that alternative, more open and complex descriptions of process are rarely forthcoming. The artifact is a rationalization of itself, one that excludes alternative forms and speaks to us thus: "I am a working, efficient, marketable machine. Knowing how I work, understanding my underlying form as the scientific principles that govern my doings, and reading my documentation (though don't be too distracted by the latter), you can reconstruct the decision-making process that made me (or rather that allowed me to make myself)."

Although the rational reconstruction of design in terms of the internal logic of the object alone is always possible and reasonable, from the perspective of design as a dynamic social process, it is wide of the mark if not wrong.[15] It is not just the difference between the Platonic circle and my poor attempt at a freehand drawing. After all, the Platonic underlying form does provide us with a means for drawing, albeit imperfectly, a circle. Rather, it is that the instrumental framing of a decision process that speaks in terms of a bounded, unique, and autonomous object and its deterministic function is destined to exclude the uncertain, the ambiguous, the nonrational, and the social as insignificant.

The limitations of this approach are revealed when we attempt to reconstruct the design of our artifact of vernacular technology, such as the chair. Oh, we can construct a likely scenario of some prehistoric chair maker, hewing a plane slab of wood with a crude stone, or perhaps a metal tool, then setting in four legs (maybe he or she was simply playing with ways of constructing stools), and so on. But this is necessarily a fabrication, made to satisfy our contemporary need for stories about causal, individual human agency and ownership. We might even pose the question in an absurd form—Who invented the chair?—as though we believed we could find a single author. How strange this question sounds.[16] Any response is bound to have the form of a myth rooted in the rationality of the artifact. Hence it might possibly be "true," but no more an accurate account of an innovative process than some children's fable known as "How the Leopard Got Its Spots," which also makes perfectly good sense if you were not in on creation or around to note and record preceding mutations.

The savant's focus on the artifact requires a fuller understanding of the way of life, the material culture, and the social, political, and economic context of early chair making—as crude as the times may have been. When we allow our thoughts to be guided by the artifact alone, we end up with a simple fable about generic toolmaking humans. Our framing doesn't contain the essentials, since the object's reach can't comprehend, in itself alone, all of the actors and agencies that make *chair*. If we try to include them, our frame loses its framing ability, and we find ourselves floating free in the crowded maze and marketplace of history.

The utilitarian's vision of who and what determines the form and function of a technology starts from the marketplace. From this perspective, it is consumers, exercising their free choice, who are the ultimate decision makers. Scholars of this sort conjure up a "free market" in which producers of goods, responding to the needs of their customers, design, develop, and fabricate their widgets and wares and set them afloat in competition with goods of a like kind manufactured by others. Consumers shop around, continuously testing the alternatives available to them. Those products that meet their needs best at an affordable, or indeed minimum, cost are the designs that thrive. In this way, you and I decide what is a good and bad design, and we are therefore responsible for the form and function of our technology.[17] In the words of a colleague, "Every society gets the technology it deserves."

Perhaps this is why some call for raising the nation's level of technological literacy: More informed consumers will be better able to articulate their needs, will harbor more technologically sophisticated desires, and will be better prepared to buy, use, and consume the high-technology machinery and systems the world is so capable of (over)producing.

When an observer of this persuasion tries to say something about the engineering design and development process, it comes out as the following kind of story:

There is a natural relationship in manufactured products between the properties of the materials that are used, the machines and processes that will be employed, and the form of the product that will result. Moreover, each of these elements must be in a constant state of change if it is to maintain dynamic equilibrium with the others. At the same time, each is not only struggling for survival against variants of its own kind but is also under challenge from other areas that seek to displace it. In a technological society that is open to change, there is, therefore, an inescapable pressure upon the manufacturer to reduce the quantity and quality of the materials and the human and synthetic energy that are being consumed without reducing the value of the product that results—in other words, to do more with less. . . .

The practice of industrial design must anticipate every eventuality in the development of products or product systems that can be manufactured and distributed economically in order to meet the physical needs as well as the psychological desires of human beings.[18]

There is a lot going on in this passage, most of it sounding scientific and lawlike, but is it meant to be taken metaphorically or literally? There is some "natural relationship" being claimed among materials, the technology of manufacturing, and the product. The designer's work takes place in the context of this natural relationship. The author doesn't spell out what this relationship is, but we can give him the benefit of the doubt and assume that it is written down somewhere, accessible to all.

But then an evolutionary metaphor is suggested: All of these elements are in a "constant state of change," yet they are in "dynamic equilibrium." Are we meant to envision varieties of the elements struggling for survival? That would leave an opening for designers to assert a more active role, perhaps by fostering mutations in the elements in random anticipation of human psychological and physical needs. Alas, this attributes too much freedom to designers working under the "inescapable pressure" of these natural relationships, pressure "to do more with less."

It is hard to see how one can instruct a designer to "anticipate every eventuality" in a way consistent with either of these two readings. The root of this predicament lies in the implied attribution of agency to the logic of machinery and the denial, or at least avoidance, of human and social agency that is reflected in the author's use of the passive voice. There are no people in this description of the process whereby we get a technology that meets our "physical needs as well as [our] psychological desires."

I do not mean to deny that one can construct useful relationships among the materials, the production processes, and the form and functioning of a product. What I do find unreasonable is the suggestion that these relationships are natural, like the law of the lever, and that they govern the design process autonomously—that designing is as passive an activity as this author suggests.

At first reading, the scenarios of the utilitarian and the savant tell two different stories about what defines the form and functioning of contemporary technology. The tension is between a materialistic conception of historical process that views consumers as an aggregate herd whose collective tastes, needs, and wants dictate the form of technology, and an idealistic conception of human progress that views each freshly discovered scientific principle as being shaped and embellished in entrepreneurial fashion for the benefit of all humankind.

Yet at another level the two stories are the same. Both are distant from the events they pretend to explain. In their attempts to provide a generic, theoretical description, they stand aloof from the sometimes chaotic workings of real people within the R&D laboratory or the engineering firm. They remain hypotheses lacking in evidence, saying very little, on the one hand, about how firms meet consumer needs or, on the other hand, about how people within the firm transform and sculpt a clever idea into a functioning artifact.

Similarly, each in its own way attributes an autonomy to the design process. Savants claim that once a scientific principle or a patent model is shown to work, then the embodiment of principle in a marketable product follows with the straightforward application of standard engineering practice, which is dictated by scientific principle again. One can envision an expert system taking over the task. Utilitarians start from the other end, where aggregates of consumers, following the laws of the social sciences, define the workings. They discount the significance of the different ways in which an expressed need could be satisfied. In a sense, the form and function of the particular technology bought into being don't matter. Or, stated differently, while there may be a particular form and function that will best meet the particular need of the marketplace, all we have to judge by is the apparent response of consumers after the fact. If their needs appear to have been met, then the product must be the best one. We can call this economic determinism—an autonomous working through of the laws of supply and demand, given the wants of the citizenry.

Both stories pretend to be "scientific," in that they offer a template or model that stands apart from chronological time. They tell a story about the way in which we get the technology we do that holds presumably for all time and for all cultures. They discount engineering design as a historically and socially contingent process.

For both stories, the object is at the center of attention, either in the form of scientific principle—the reality behind the appearances according to the savant—or in the form of the needs of the aggregate of consumers—the reality behind the appearances according to the utilitarian. It is not a matter of chance or poor writing style that the passive voice predominates in both accounts. Both, in these ways, show symptoms of what I would label an object-world view of a social process.

A Broader Vision

The object-world way of seeing cannot capture the process of designing. There is no science of design process in the way participants themselves understand that term. This is not to say that the process is irrational, that a story can't be developed and told that makes sense, or that one cannot, on the basis of this story, infer improvements in the process. It *is* to claim that to be "scientific" about the study of design process one must admit the possibility that the object—as either physical principle or economic necessity—is only part of the picture, and a very fuzzy part at that.[19] If we want to understand design process, we must remain sensitive to the full breadth and depth of social context and historical setting. Heather Lechtman and Arthur Steinberg are archeologists who have studied the shaping of ancient technology and, reciprocally, the ways in which technology has helped define cultural patterns and organization. They have posed the question in the following terms:

One of the issues we see as uppermost in understanding the nature of technology within cultural settings is the integration of any technology or system of technologies with the cultural matrix in which it is manifest. The question of integration is concerned with the extent to which any technology has a life and a style of its own, bringing with it a set of inherent properties or characteristics that are inescapable and independent of specific cultural milieus. Some of these characteristics may be grounded in the physical properties of matter itself; others may relate to the energy sources necessary for the technical manipulations taking place. On the social structural level, there is the difficult but crucial question of whether specific technologies require and predict certain organizational forms or whether the social organization of technology derives from that of other cultural subsystems, such as kinship or religion. Within the range of expressions any technology may take, how much of a given manifestation is predicted by its materials and energy systems and how much by organizational modes, value systems, mythologies, or science?[20]

How do we do this? How do we explore a cultural matrix, investigate a specific cultural milieu? The task is particularly challenging for those, like Lechtman and Steinberg, who want to understand social processes in another culture from a time long past.

Take our anonymous, ordinary chair again as an example: There are no written documents explicitly describing its invention, no drawings, no entrepreneur's life story that has come to us in ballad much less prose. *Chair maker,* like the *chair* itself, remains anonymous. Indeed, one can legitimately ask whether there were specialists in the business at all or whether chaircraft was part of every householder's repertoire of skills. Our interest in the object as a thing in itself confronts the murky canvas of *chair* in its historical context. How can we reconstruct the emergence of the object out of the fabric of community life, of ritual, of work and repast—a milieu, too, of *chair* as symbol, metaphor, or myth as well as functional convenience? To do this we must unfocus, start with a broad canvas, hold suspect the categories and relations we unconsciously accept today, and seek, like Lechtman and Steinberg, among historical remnants—painted, sculpted, written, or wasted, whatever is left to be excavated—for evidence of relations in the making and using of a chair.

For modern technologies, matters are different, yet the prescription still holds. We are surrounded by the artifacts we may wish to research, and their sources are apparently well defined. Today we have, or appear to have, closure around our technological productions. Our x-ray machines come in a box with a label warning us about special procedures we must institute to use them safely. We can identify specific places where design and production get done—some ringed by chain-link fences or at least there is a book visitors must sign and a badge they must display, marking them as outsiders in order to gain entry. We can name the enterprises that produce these things, corporations with identifiable executives managing, scientists researching, engineers designing and developing, and others marketing and selling. We can easily locate the artifact as a thing in itself and enclose it within its own context. Here is where one goes to study contemporary engineering design.

Donald MacKenzie and Graham Spinardi's analysis of the design and development of ballistic missile guidance systems is exemplary in revealing the complexity of decision-making processes when distinctions between "technology" and "politics" are difficult to maintain and when "bureaucratic politics" intervenes between rational science and market needs.[21] This work comes close to the spirit and method of the studies I will report here.[22]

But while case studies such as this reveal an often uncertain and ambiguous process, they also remain somewhat constrained. The after-the-fact recollections of participants in any process are going to be shaped by object-world thinking, that is, by rationalizations of why event B followed A. Confronted with this, the historian's usual tools may not suffice. We would do best, I will argue, to enter the firm in "real time," to break free of the hold of instrumental rationality and see what goes on in the day-to-day, often dreary and mundane, but frequently exciting, process of design.

Contents

Thus we shall go inside the firm to observe the process by which we get the forms of technology we do. We want to test the notion of autonomy of technique insofar as this excludes the choices available to designers, and thereby to you and me. I am not interested in positing better alternatives. That would be presumptuous. Nor am I concerned with the technologies that occupy current news space, such as nuclear power or office computers. Rather, I will assume that, at some level of abstraction, the design process is the same everywhere. If the form of a photovoltaic module is indeed dictated by "trade-offs" among strictly instrumental factors, I ought to encounter the same situation in the design of x-ray machines and photoprint processing systems.

My working hypothesis is that the process is not autonomous, that there is more to it than the dressing up of a scientific principle, more than the hidden-handed evolution of optimum technique to meet human needs, and more than the playing out of the bureaucratic "interests" of participants seeking power, security, or prestige. In the affirmative this hypothesis takes this form: *Designing is a social process.* Executive mandate, scientific law, marketplace needs—all are ingredients of the design process, but more fundamental are the norms and practices of the subculture of the firm where the object serves as icon.

My plan is to start with a close-up look at the object of design as seen and worked upon by participants. In chapter 3 I relate some stories about participants in design working within object worlds. Then in chapter 4 I construct a cosmology of participants' object-world thinking, showing how the object structures their thoughts, beliefs, and practices, not just when they are working within object worlds, but throughout the totality of life within the firm. This moves me away from the object toward the organization and infrastructure (the "ecology") of the firm as subculture. Now, with a setting, a cosmology, and some sense of the three projects, I turn to narratives of process. Here is where the object loses its hold. In the simplest terms, design is the intersection of different object worlds. No one dictates the form of the artifact. Hence design is best seen as a social process of negotiation and consensus, a consensus somewhat awkwardly expressed in the final product.

The last chapter brings us back into history and describes designing as a historically contingent process, like any other human activity. Closure on the three projects reopens, indeed forcefully brings to the fore, the question of context and of still other boundaries. But our immediate next step is to focus on method.

2
Engineering Observations

In the old days, anthropologists studied others, far afield, but recently both anthropologists and sociologists have taken to doing fieldwork in their own backyard, studying indigenous groups of the aged, the poor, the police, medical technicians, and the like. Still, to label a gathering of engineers a *subculture* sounds a bit pretentious. After all, life inside a firm is not bounded or complete in the same way as it is for the Navaho or Walbiri, nor does it display the distinctive features of a gang of East Side punks or New Age survivalists. On the contrary, the employees of Solaray, Photoquik, and Amxray are ordinary citizens like you and me, and certainly we would count them as members of our common national culture.[1]

What makes the corporate activity of designing a subculture is that those engaged in the activity, in going about their work, espouse a common goal and share particular priorities, beliefs, and values; they call upon special forms and bodies of knowledge and use a sometimes jargon-laden, instrumental language; they interact both within the firm and with others on the outside in distinct ways; and their identity and survival depend, in large measure, upon the viability of the group and its productions. In this sense they are not like you and me. They, as a community working up their projects, speaking their own language, on their own turf, are different from outsiders. You sense this when you enter the door: Going from one firm to another is a bit like going from one country to another.[2]

There is nothing pioneering in using the term *subculture* to describe the thought and practices of employees in an engineering firm. Anthropologists and sociologists at graduate schools of business administration study "Organizational Culture," a subject that is now part of the mainstream of the MBA curriculum.[3] Their numerous studies, however, rarely focus on the work of participants engaged in design, nor do they attempt, as I shall do here, to explain the significance and role of instrumental reasoning and engineering technique, not only in that work, but also in the life of the firm as a whole.[4]

Anthropologists see tools and technique as but one of several dimensions spanning the social fabric of a community.[5] In their attempts to understand, describe, and explain the *other*, they study, in addition to material culture, their beliefs, ways of thought, language, rituals, and symbols; they observe how people interact with the environment and how they marshal resources to ensure their well-being; they record kinship patterns and learn how the elders of the community indoctrinate the young through the passages and stages of life's cycle; they listen to stories about traditions and history and infer a conception of place in time on a grander scale. These threads woven together describe culture, whereas no one thread alone reveals very much about the whole.

So, too, in the contemporary engineering firm, in our attempt to explicate the design process, we can study participants' modes of thought, systems of communication, and beliefs about what constitutes good design practice. We can take stock of the infrastructure that supports designing—the whims and needs of the marketplace as well as the resources and facilities available at a price from other firms. We can study the organization of a design project, informal as well as formal, and how new recruits learn their place in the operations of the firm and, in particular, how they learn to use the sophisticated tools and techniques that are crucial to product design and development. And we can listen to, as they do, stories about triumphs and tragedies in bringing the product to market. Again, each thread helps fix the pattern of the whole, whereas no single line of investigation reveals very much about designing.

For example, management studies have noted the importance of the spatial organization of a building to communication and information exchange within and among different groups in a firm. Yet floor plans alone do not fix the ambience, character, and quality of group members' interactions. Another firm could occupy the same site (after extensive renovation to exorcise the spirits of the previous occupant) and carry on in distinctly different ways. Or the firm itself might move into a building of a different plan, and that change need not significantly change the ways in which employees see each other, exchange information, or negotiate their differences.

Picking up another thread, we note that the special character of a culture has something to do with the skills and experiences of individual participants. For example, a pioneering entrepreneur might set the stage for the conduct of business and periodically rally employees to a common cause.[6] Although the chief's influence can extend well beyond his or her retirement, we must be careful not to attribute more to the heroic exploits and expertise of any individual than is warranted. People come and go, are hired, fired, shifted from one group to another, move on to a better job with a competitor, or retire, yet the subculture reaffirms and retains its special character.

The dynamic nature of life within the firm, as displayed in the surprisingly frequent moves from one place to another and in the shifting of people in, out, and about, suggests that use of the term *culture* is inappropriate. How can we speak of enduring patterns of structure, norms, roles, and the like in the midst of such change? Yet despite this flux, we find an anthropological perspective useful to our understanding, and hence justified. Indeed, the flux itself demands an explanation, a task best addressed by thinking broadly in terms of culture.[7]

In this study we posit that technology is an essential and primary dimension of life within the firm, one that determines its special character. Technology figures largely in a broad range of abstract and practical experiences: in designers' use of instruments and tools, in their analyses of the performance of objects, in their drawing and sketching, planning and budgeting, in the way in which they organize their work, and in the metaphors they use. Objects are continually at hand as a focus of thoughts or a topic of discourse.

The significance of technology in our lives is often unrecognized. Yet as we move in and out of our different worlds—home, work, play, on the road—technique constrains and guides our associations in some settings to such a degree that it patterns our beliefs as well as our actions, shaping our values on a grander scale and limiting our aspirations and expectations. At the same time, though, technique is continually shaped and reshaped as we put it to use.

Consider automobile commuters as an example of an active subculture whose members engage in a common enterprise and share a set of norms and practices that is strongly influenced by the technology with which they work. Commuting, as most of us know, requires the competent use of technology, and this both defines the nature of our experiences with others and serves as a milieu for creative expression.

Looking, however, at the technology of commuting by itself alone as presented in advertising media tells us little about the experience of commuting. The macho machine high-jumping over dusty western foot-hills and the sleek black carriage discharging its perfectly dressed passengers at someone's colonnade of a front door by the light of a full moon are not commuting. Nor are the roadways displayed in commercial images—the aerially viewed, six-lane ribbons of highway or the static, Victorian intersections in Norman Rockwell settings—adequate representations of the experiences we have traveling our favored route to and from work. Automobile and roadway are indeed integral technical ingredients of commuting, but their popular representations in the media tell us little about what it takes to participate fully in the life of a commuting subculture.[8]

What distinguishes the real experience of commuting from that suggested by the glossy pictures and what justifies the label *subculture* is that we commute *with others*, and our success at it depends upon all the doings of those others—upon the frequent instantaneous associations we form and the negotiations over rights of way and other common maneuvers we continually carry on with those around us. Commuting, in short, is a social process.

From this perspective, the automobile and the roadway appear not as shining, efficient, static artifacts to be occasionally engaged at our leisure but rather as the familiar clothes we put on each morning. Here is an example of daily ritual and obeisance: As we climb into our car and back out of the drive, we become full participants in the subculture.

We can even speak of a commuting cosmology—a shared system of beliefs about what constitutes a correct maneuver or illegitimate practice. These need not be articulated but, like vernacular knowledge, constitute the craft knowledge of participants. The formal rules one finds in the manuals of the Department of Motor Vehicles do not do justice to the richness of that craft. Think of the rules governing behavior at a rotary, and then think of how you actually survive one. Formal rules are necessary, of course, not only to maintain some semblance of order but as material for testing prospective members before they are allowed out onto the road; the initiates in most cases are youths in their teens who see the event as a watershed on their way into the world.

While passing the exam provides entry, a good bit of tacit knowledge is necessary to participate fully in a commuting subculture. To a competent driver, the apparent distinction between a legitimate move and wrongheaded behavior is often marked by the presence or absence of the briefest of signals—a suggestion of a veering to the left or a flicker of a brake light. There are nuances here: The slightest difference in acceleration, speed, or rate of turn can separate the ill-intentioned maneuver of the irate driver from the acceptable pass-on-the-right or cut-in-front-of move of an accommodating neighbor. A colleague once defined a microsecond as the interval between the light in front of you changing from red to green and the person in back of you hitting his or her horn.

Within this environment we have distinct ways of communicating; signals, gestures, and, at times, not looking will serve, and although we may often compete and curse, we share the common goal of getting to work on time, sane and undisheveled.

There are different varieties of commuting subcultures. In part this can be attributed to place: One person sticks to country roads and winding suburban streets while a neighbor finds the nearest turnpike entrance. In part it can be attributed to differences in technology: The large eight-cylinder vehicles gassing through Los Angeles over an expansive network of freeways define a different commuting subculture from that of the Deux Cheveaux weaving across Paris. Of course being French, American, or Mexican matters too. We might speak, then, of different subcultures of different peoples, apparatuses, resources, and ways of carrying on, and identify each with a specific place and time.

We must also acknowledge, however, that all of these technologically based associations are the same in a sense. Commuters in Mexico City have a lot in common with commuters in Los Angeles, Paris, or even London. From this perspective we can see commuting as a universal life experience. We might then imagine, continuing with this fantasy, a collection of anthropological studies of commuting subcultures and comparative analyses of similarities and differences. Might there even be a deep structure to commuting? Might we be able to construct rules of transformation that would enable us to see the maneuvers of a competent Londoner driving on the "wrong" side of the street as being the same at some fundamental level as those of a Parisian on the way to work?

In sum, the commuting experience is social but mitigated by and transacted by means of a technology that envelops (literally) the members of the subculture. The vehicles, the rules of the road, and the infrastructure of the urban landscape both enable and constrain the ways its members perform, setting a framework for thought and decisive action.

This example suggests why workers in an engineering firm can be called members of a subculture. Within the enterprise, participants in design follow different routes, work at different jobs, and have different responsibilities and interests, yet they are engaged in a common task and share a common goal. They shape to their own purposes a knowledge base that is essential to design. They communicate using a language tuned to the specialty of the enterprise. They conduct their business according to an often implicit but sometimes explicitly articulated system of rules and norms. And members of a firm are disciplined in thought and action, not just by formal and informal organizational rules and norms, but by the objects—the physical hardware, scientific concepts and laws, rules of thumb, design specifications, codes, cost comparisons, and milestone charts—that they must engage, reconstruct, and manipulate in the course of design. These are the threads spanning the work of design that justify the label *subculture* and invite study in the field.

My attempts to trace these threads and construct a viable description of design process will be based on a study of participants in three design projects at three different firms, labeled here Solaray, Photoquik, and Amxray. The duration and intensity of my observation and participation differed among the three: My fieldwork at Solaray stretched out over two years. I was at the firm full-time during two summers and visited one or two days a week during the academic year. I observed the project at Amxray over a nine-month period, culminating with the demonstration of the prototype. I was on site full-time during the summer and at least once a week during the preceding spring semester and subsequent fall semester. My contact with Photoquik was less frequent—nominally once a month over a period of a full year. I also had access to raw video recordings of meetings and interviews made by a group within the firm who had engaged my services as a consultant and were exploring the value of this medium as a communications tool in the design process itself.

The extent of my participation as engineer also varied from one firm to another. At Solaray, I contributed to the structural design of the frames for the photovoltaic modules and developed methods for predicting the performance of systems in the field. At Amxray my participation as engineer was less substantial, although my perspective on technical issues was solicited on occasion. At Photoquik I advised the team doing the video recording. Rarely did participants in Photoquik's design task solicit my technical advice, although they did not hesitate to explain what they were doing in common engineering terms.

The study of a culture up close raises a host of interesting methodological questions. Participant-observation of a community, especially when the natives are not very different from the observer, calls into question the classic differentiation between subject and object. Anthropologists and sociologists are well aware of the difficulties inherent in ethnographic fieldwork. I turn now to some particular questions about access to the process of design, about the means available for taking data, and about reading and interpreting all that I saw and recorded, addressing these methodological questions within the context of observations made within the three firms.

Solaray

Not everyone around Boston funnels in toward the city center to work. Some ten miles out, ringing the inner suburbs, runs America's Technology Highway, Route 128. Here are the high-technology firms that have spun off from university research centers, larger corporations, and federal laboratories. Some of these firms occupy multiple buildings scattered over acres of manicured lawn, while others find their niche in groups within buildings nested in outcroppings of granite and macadam. The photovoltaics firm Solaray is located in one such building.

Driving there in the morning gives me a sense of freedom. I motor against the main thrust of the traffic, heading west for the country. I could as well be going on a picnic.

My initial sense of the place is chaos.[9] I am given a tour, shown the conference room that doubles as a library, meet the president, and encounter a Coke machine, different group leaders, other visitors, a sandwich truck, photocopying machines, technicians in the lab, and others. There are people doing paperwork, typing at computer terminals, drawing at a board, and sorting mail; small groups are talking and laughing. I come away confused, my notes disjointed. Recorded experiences remain unconnected. I don't know the lab from the shop, nor can I distinguish a technician from a project manager.

My first encounters with the people I intend to study are awkward; they require too much explaining. There are reservations, ambiguous questions, acts of showmanship and probing. Management is initially concerned that I will be a nuisance. Engineers are curious about what

I am doing there. I am, in point of fact, no stranger to their trade. I had already worked with a group at MIT's Lincoln Laboratory on the design, fabrication, and field testing of photovoltaic, solar energy systems intended as demonstrations of this newly emerging technology. This experience had been critical to my gaining access to Solaray in the first place.

After a month or so, once I have explained my mission and they have fixed a place for me in their environment, I become an insider of sorts, although the distinctive badge I wear marks me as different. Even now my presence engenders a more constrained and studied ambience than what I suspect exists in gatherings of full-time participants. There is no fooling myself: No matter how established my participation, there is always a critical distinction between me and them—my welfare is not as directly tied up as is theirs with the fate of the enterprise or the success of the design project that engages them. There are, at first glance, advantages to my status, for I am perceived as independent, as someone to whom participants can talk and explain the intricacies of the problems they face, and they know they can seek in me acknowledgment, constructive criticism, or just shared laughter.

With time I gain some bearings. I am able to remember people's names, although I no longer notice many faces. I can identify participants meeting informally in groups of two or three and can pick up on the intent of their gathering without interrupting their conversation. I search for meaning in the casual conversation and try to distinguish when laughter is critical to the flow of negotiations. I learn who are the old-timers, who the newcomers. I sense hierarchy and how it doesn't necessarily correspond to what is shown on the formal organization chart. I become better attuned to the scope and depths of the different design projects that claim people's attention.

Some of these are recent undertakings—the development of a large photovoltaic module for residential applications, for example. Others date back to the firm's first attempts at putting their product into the field; a photovoltaic-powered desalination plant installed three years ago in Saudi Arabia is so ancient a project that it provides the stuff of heroic tales of problems overcome.

Today when I visit, a relaxed ambience appears to pervade the firm. There is going to be a group meeting of Marketing and Systems Engineering. I, as usual, attend and take copious notes.

Beth is going to be late. Should they go ahead without her, pick up on her piece of the proposal when she arrives, assuming she does arrive?

Brad, the manager of the group, pulls out a styrofoam coffee cup, half filled with change, and pushes it toward the middle of the table.

Brad   *We might have to up the late entry fee to a full dollar if this keeps up. Where's Tom? Oh, there you are; you're lucky I didn't get here on time. Have you seen Beth?*

Tom hasn't seen Beth. Brad decides to go ahead without her and distributes a photocopy of his outline of the proposal that is the subject of this meeting.

Brad   *Sorry about the teeth marks down there near the bottom line; our new dog got hold of my printer output at around 1:00 this morning. I didn't have time to reprocess. OK. Everyone got one? We've got but a short week to get this thing together and off to Riyadh. We don't want to be buttoning this up riding the Federal Express.*

The request for proposals for the design, fabrication, and installation of highway lighting systems in Saudi Arabia had arrived just a month ago. Brad had learned earlier that some significant work in photovoltaic-powered lighting was on the Saudis' wish list but that the details of the design specifications were not in hand until just a few weeks earlier.

Brad   *Ed, have you been able to get anything going with Lorings Saudi?*

Lorings is a large contractor capable of doing the heavy work of assembling the photovoltaic array support structure and designing and constructing the auxiliary buildings required to house the batteries and controls.

Brad   *We've got to have some kind of piece of an agreement with them to put them in, even if we can't be too specific on costs and the extent of facilities. Do we know what kind of space we need for batteries? Where is Beth? Tom, do you know if she's got an estimate of how much storage [energy] we will need on that tunnel lighting system? That's the big unit on this job.*

Tom   *She had worked up a number, but she was waiting for Ed to come back with a better idea of what it would take to power the lights. What was. . . ? Oh yeah. Got hold of Lorings Friday. They are interested. I told them this was a job about three times the scale of the desalination plant. The only problem they might have is scheduling us in. They said to go ahead, put them in; their engineers would get off a cost estimate in a day—rough but conservative so they're not too worried. We should have that today or tomorrow.*

Ed   *How about their ditchdiggers? They able to get skilled enough workers in the middle of the desert?*

Tom   *The way the wind whips around out there sometimes its not a bad idea to have a few ditches around.*

Ed   *I'll take it over yesterday's blizzard . . . or was it just a heavy flurry? I'm going back to Phoenix.*

Tom *Did you see that show on Channel 2 last night? All those Saudi princes driving around in their Caddies? What the hell do they need tunnel lighting for? Why don't they just level the mountain?*

Brad *Ed . . . the lights?*

Ed *They're looking good. Cellvue is eager to put their product in a photovoltaic system. This would be a big project for them. They see this as an excellent opportunity for showing off their controls. They're talking about modulating the intensity, depending upon intensity of traffic, time of day. . . . I've been trying to get an estimate out of them on costs, trying to restrain them on their fancy software, but I do think they have some good ideas. Mike is going to stop by this afternoon with three options. I'm confident we can go with one of those and the dollars will be reasonable.*

Tom *I did my sizing [of the photovoltaic array]. If Cellvue comes in with 10–20 percent less energy requirements, all to the better.*

Brad *That may be good enough for the proposal, but we want to be sure we leave ourselves some room to move . . . more conservative.*

Tom *I've got the array all sized, 10 kilowatts, including the support structure, wiring, diodes, switches, and I've got a pretty good idea of what the controls are going to look like.*

Ed *Hey, Brad, I heard that Solarpack is trying to pull some strings through their big oil contacts. Haven't we got some access?*

Brad *That's Solarpack's way of doing business. They will try every trick in the book, and some new ones to be number one in Saudi. I don't want us to go down that road. John, what's this shaping up to be in total dollars?*

John *With what Ed and Tom have given me, and using the four-day storage number, I think we can come in at a competitive price. We've gotten the go-ahead to use the inside price on the modules. We're going to try to make this a job for the residential module. That's going out on a limb, but production is confident they can get the bugs out of the fabrication process in a few months' time. They are still waiting on our final design of how the "res module" is to be wired, where the junction boxes go.*

Brad *How do array costs change if we have to go back to the current product, our 30-watt module? The support structure cost goes up too.*

John *I think we will still be alright but I would like to bid with both as alternatives.*

Brad *That's not possible. Oh, here's Beth.*

I sit there in the midst of talk about the weather, the group leader's new dog, Beth's tardiness, Saudi Arabians driving Cadillacs, competitors pulling strings behind the scenes, and I ask myself, in what way is this *data*; is this a significant event worthy of record if indeed it is data at all?

The question still pops into mind, even after I have been around for almost a full year. Do I have access, or is the really significant data to be found behind other doors? Certainly I am bound to miss some important events. Even if I were at the firm all hours of the work week, I could not possibly be everywhere at once maintaining contact with all those who have a legitimate interest in and responsibility for moving the design along. Many are dispersed about the firm and affiliated with different organizational units; a key player—a supplier, for example— may even be located far off in another country. I am bound to miss a critical conversation, an informal but decisive meeting, or a significant step in the redrawing of plans that might occur elsewhere at another time—for example, in the negotiations over the phone with the illumination subcontractor, or in the prototype testing of the battery controller down in the lab.

Some meetings are more important than others and are known to be so before the fact as revealed by a formal agenda. I judge their potential significance by noting who is scheduled to be in charge and recalling their authority and responsibilities. Yet a casual meeting can become formal in just a brief instant, and important things can be established in a hallway tête-à-tête. Clearly, then, here is the challenge of fieldwork: to choose judiciously, or hope to happen upon, occasions for observation, knowing that I cannot be everywhere at once recording all potential data, leaving aside for the moment the question of what to record when I am present. I take some consolation in knowing that no one participant, not even a group or task leader, has complete knowledge of the myriad of events and exchanges that contribute to the ongoing design process.

I strive to be around enough so that I can claim to be in touch with the pulse of the project and to keep pace with the ups and downs of all participants as they go about their daily work. Generally I'm on site two or three days a week. At times the pace slows down to a point where the design barely seems to budge over the course of a month. At other times intense activity bursts out over the course of a single day and continues into the night. If I've missed the action, I glean what I can from informal interviews with participants the next day. In this way I stay engaged to the extent that I sense continuity in day-to-day events and can even anticipate the reaction of one participant to another's provocation. The challenge is to remain alert to mark and record the potentially significant evocation that may emerge in ordinary conversation. This runs at odds with becoming comfortable and familiar in the midst of process.

I cannot afford to become too comfortable; I have to keep thinking about how my presence affects the discourse of participants and, hence, what I record in my field notes. I wonder, too, how my intervention, as passive as it may be viewed, might possibly affect the process of design and the design itself.

Other problems of orientation and observation stem from my un-awareness of the traditions of Solaray. I almost immediately learn that the proposed photovoltaic system for Saudi Arabia is a variant, albeit a significant one, of preexisting systems. I speculate that the design process I observe now is relatively tame and ordinary compared to what went on in the earlier days. The inertia of design is well known.

But this just begs the question. The entrepreneurial history of the firm is not available to me as a living process.[10] Furthermore, the design process that I observe shows little of the inertia or the autonomy alluded to in the first chapter. Instead I see a dynamic process, where the past does matter but is continually challenged and forced to justify itself in terms of the present. Even the myths about precedents are construed to speak to today's concerns.

I take notes at group meetings and design reviews and at smaller but still formally arranged meetings of twos and threes. I do my editing in situ, recording in writing—better seen as a scrawl when the discourse gets intense—specific language and gestures, sketches on the board, comings and goings, laughter and yawns, the spoken word as uttered. I summarize and expand upon my rough notes at the end of the day or early the next morning. I rely on memory but find I can recall the sequence of a conversation, its rhythm and meaning, prompted by my notes. There is an editing process going on here, an evaluation, rooted in a framing of the situation, that I bring to the event and that shapes my reconstructions.

[Field data] are not like dollar bills found on a sidewalk and stealthily tucked away in our pockets for later use. [They are] constructed from talk and action.[11]

The dialogue presented above, then, should not be taken as a literal transcript. It is a fabrication, an account built up from my field notes. It is factual in the sense that Beth was indeed late, Brad did report on his dog, Ed and Tom did banter about ditchdiggers and Saudi princes while the group worked up a response to the request for proposals for the design, fabrication, and installation of several photovoltaic-powered highway lighting systems for Saudi Arabia. The dialogue is best under-stood as a *paraphrasing* of what I observed, heard people saying, and saw them doing during the two hours they were gathered together around the one table in the room. Think of it as an *encoding* into a new dialogue spanning a few pages of text (which you read in a minute) of my *decoding* of the events and exchanges of the meeting (which I worked up in more than a few hours). It is an *account* of what transpired, one (and not the only one) whose veracity can be tested by you as reader as well as by the participants themselves.[12]

My field notes would have benefited if they had been supplemented by a tape recording. Indeed, some investigators do focus intensely on bits of conversation precisely recorded in the field and draw out of their analysis of this microevent rich information and insights into the patterns of belief and the meaning of specific practices of the parties to the conversation.[13] A knowledge of context is just as essential in this mode of investigation as it is in my more gestaltish reconstructions.

I did experiment with recording on site and, like others before me, found that the process of transcription is not distortion free. The audio recording of a multivocal discourse, likely to be the most critical part of any meeting, can be difficult to untangle and set out linearly on a piece of paper. The event often calls for an evaluation that runs counter to serialization. Serialization can seriously distort a record that, although sounding like a cacophonic performance of a piece for woodwinds and strings played by a class of fourth graders, may very well have enduring meaning to the players, as unharmonious as it may sound. As a proud parent who is a participant-observer, you too might hear a single movement despite what others say.

A tape recording also misses gestures, winks and nods, yawns, sips and nibblings.[14] These micromovements can be important. Like the details of the brush strokes in an Van Gogh or Monet, they enhance the color, shading, and nuance of a scene and ultimately fix its distinctive meaning. There is also the mundane but nontrivial problem of the oversupply of "information": Ought the shuffling and noise on the tape be transcribed? Then again, how does one decide when to turn the tape on and when to turn it off? Not all the important exchanges occur during the official hours of the meeting; the coffee break can be the most critical time—a time when an irksome question gets answered or a fiat is proclaimed one-on-one.

While a recording can serve well if "read" in conjunction with notes, tape recording alone is a restricted method of analysis. It is not a neutral, transparent instrument but rather entrains a particular way of framing events conditioned by the instrument's qualities and the acoustic character of the space in which events are recorded, as well as by all of what goes unrecorded or is only faintly suggested. It is a filter, like a pair of sunglasses, that allows certain questions and disallows others. "What

exactly did John say?" is allowed. "What was Michael's unspoken, immediate reaction?" is not allowed. Nor are "What was Ed's doodling all
about?" "Why did John have so much trouble getting his sketch right"
"What shifting coalitions?"

And then there is video. . . .

Photoquik

The second firm wanted to experiment with the use of video as a
medium for exchanging information in the design process. Photoquik
Corporation is much bigger than Solaray. It has divisions in the western
United States, in New England, and in West Germany and the United
Kingdom as well. Like so many large American firms, their lead in
automated color photoprinting technology has been eroded by foreign
competition. Pressed more and more each year, and having yielded
some market niches to the competition, Photoquik struggles to keep
hold of the high-volume end of the photoprinting market. Their machines are still favored by job shops. Their most recent product line is
pressing the state of the art in color development and printing along
all technical fronts—mechanical and electrical as well as chemical.

The first indications of a problem in the quality of the printed image
appeared some six months ago in the course of the design and development of a new product tagged *Atlas*. The problem initially was not
considered serious: Engineers conjectured that it could have been
caused by any one of a number of design features yet to be fully detailed
and proof tested. Expectations were that when all the pieces were put
together and running, the problem would go away. It did not.

The potential problem is most easily stated in terms of what was *not*
observed: "Dropout," as they call it, is a failure of the process to fix the
printed image at hardly visible spots on the print paper. For an observer
not looking at the "microstructure" of the image, dropout points are
often difficult to pinpoint. To do so takes an effort akin to forcing
oneself to identify the individual pixels that make up an image on a
video monitor. It was not clear when the problem was first noted, or
how serious it was, if at all. But recently, engineers at Photoquik's
Colorado Division were also talking about dropout in their product line.

Sergio, a project manager of the Advanced Technology Division, was
not surprised when his chief asked him to put everything else aside and
lead a team to help the Atlas product development group rid themselves
of this problem. It wasn't the first time he had participated in this kind

of fire drill, but the chief's words suggested that a more substantial effort than usual would be required. He was told to call together everyone in the Manchester facility who had some experience with this kind of defect and the technology that was its probable cause, to review the status of Atlas, and then to come up with an option that could then be pursued as a "fix." The chief suggested that the problem was considered sufficiently serious that significant resources would be committed to its solution.

I have notes from the first meeting called to discuss the dozen or so redesign options that Sergio had rounded up over the previous month in talking with his specialists and product development people. The meeting was intended to narrow down the options that would be pursued further from fourteen to two or three. This was my first day at Photoquik.

My rough notes are barely decipherable. Scattered over ten pages I observe some micropoints, pixels of information, out of which I struggle to construct a coherent pattern, a true image of the conduct of the meeting. Most of my scratchings are about language—words and phrases that were new to me. That's not quite right. The words were the stuff of everyday common discourse, but the way in which they were being used—their nuances and implicit meanings, some hints of jargon—all masked any continuity of expression to which they, the attendees, seem to be so effortlessly attuned. I was having problems composing a coherent record of the event out of my notes. I thought I had understood the main features of Sergio's presentation of the options, but there was much I did not understand. How much of this was attributable to my lack of understanding of the technology, the science of photoprocessing, and how much to my ignorance of context, alliances, and participants' interests and histories?

I turn to the video record made of the meeting. Replaying the tape helps me immeasurably. What I see on the monitor before me seems as ragged and incoherent as my rough notes, but in a different way.

My notes bring back a vision of a fully three-dimensional room. It was too narrow for the meeting. A long table filled up most of the floor space. White boards lined two adjacent sides. On the other long wall was an expanse of windows. A table with coffee and doughnuts at the rear further contributed to the congestion. The crowding of video equipment and the two people running this machinery into the space between the coffee and the entrance to the room also contributed.

The video recording shows none of this. The field of view and mobility of the single camera operator are tightly constrained. I see only a two-dimensional framed surface, practically empty except for Sergio and his charts:

Sergio   *Where were we? Oh yeah, number four. Number four is an air wash. The idea is to use an air jet to keep the paper against the bed and, at the same time, distribute the flow uniformly.*

The presentation and discussion of options is in high gear. Sergio has a flip chart, one chart for each option. Some are but slight variations on another.

Sergio   *This is a bit of the untried but we want to check it out, see how it sets off against the rest.*

I hear the murmurings of a chorus, offstage, stop the tape, rewind to listen again.

Will it blow the emulsion off the bed? Where will we position the supply? What pressures are we talking about?

I hear mostly questions. Some of them are antagonistic; most are constructive and well intended. I am aware of the others in the video through Sergio's responses. Often the camera operator zooms in for a close-up of Sergio, then backs off for a far view; but in even in the far shot, Sergio, addressing the camera, predominates. Only rarely does the image include a full frontal view of any of the others in monologue. More often they appear as backs of heads or stunted profiles that point away from their remarks to the front of the room. The video does the framing, sets the scene. An interrupted, disjointed monologue is the result.

Sergio   *I think we can fit it in here. It may be tight, but if we can get Atlas to move this feed roller.*

I hear of contingencies and uncertainties touched by ambiguity. But that's OK. This meeting is meant to get the options out onto the table. Leave things open now. Decide about narrowing them down to just a few later. Sergio has a methodology he wants to try out for that purpose.

Sergio   *Number five. Number five is an electromagnetic excitation of the bed; high frequency, E&M . . . Fritz has been playing around with this in his lab and he's gotten some interesting results.*

Again the chorus:

What are the power requirements? You know we don't have much left in the way of an energy budget. And what kind of power conditioning do you need Fritz to supply your gizmo?

Fritz, off camera, responds, but not precisely. (I know it is Fritz from rummaging through my notes.) There are further questions about fit, about how the thing works anyway, who will supply the transducer, what sorts of control options are there, and so on.

Sergio *Number six is the old standby. We did it on Venus. So they want us to consider it for Atlas. But I've got a problem here. That roller was a cludge of a fix and I don't think we can fit it in here, scale it down, and still have it work. JC, it's tight in there. Have you seen the way they've packed things in on the paper supply module?*

Hans *Sergio, when are we going to get one of their prototypes to fool around with? Our tentative schedule calls for some hardware mock-up and testing in two months.*

Hans is the only other person in the room budgeted for the task to date. He has more experience with, and knowledge about, the intricacies of the chemical aspects of the print process than do most people in the plant. He is a good complement to Sergio, who is a mechanical engineer by background.

Sergio *I've put a word in to the chief and talked to [?] over in Atlas. They've promised me we will have the machinery when we need it. But I've got to tie them down on that. They're really getting into a scramble mode over there. The way we've got to go is to assume it will be there when we want it. I'm getting a technician, or at least going to make the strongest case I can for one of their experienced people.*

(Hans) *That's a lot to ask of them.*

The meeting goes on another hour past the coffee break. I fast forward the tape.

Sergio *That's all we could come up with. I know it's late, so we're going to have to get together again . . . and do the Pugh Method. I sent you all a copy of that, didn't I?*

PT *Serge . . . I think we have another option. It's so simple it doesn't come to mind.* [PT is the most experienced engineer in the room.]

Sergio *What's that, PT?*

PT *We can design that entry baffle to force the paper, hold the paper onto the bed. We have the program to do the analysis. We used it on Jupiter.*

George *Can we really fool around with that?*

Marco *That can't do it all. How much can a baffle gain you over the entire span?*

Hans *On the other hand, it's a quick, cheap fix. . . . I think we should add it to our list.*

Sergio *PT, you say they used it on Jupiter? And it worked to give them some improvement?*

PT   *I don't know how much. I just know they used it. They saw our model and our output so they tried it.*

Sergio   *OK. But let's not call it another option. Since it can work with most of the other things we have been talking about, let's just note it, and by that I mean we follow up on it, take a look at how it might help us. PT, you have time to do some modeling?*

PT   *Not much. But I will make some, for you Serge.*

Sergio   *Thanks. OK. When do we do this again? It's fun, huh? That's the toughest part of this act—getting all you guys together in one room. Tell you what, I'll have Michelle do the scheduling but tentatively. How does next Wednesday look?*

In principle, a video record contains more information than an audio record alone. Certainly, anyone who has responsibility for digitizing a recording and storing the bits of information in a computer knows that video requires much more memory than an audio recording of the same duration. But video-audio data of sight and sound together are qualitatively different from a simple sum of the two. How informative is it to watch a video record without sound? How does the character of audio data change when heard alone and then in conjunction with the video? Rather than speak in terms of "more (or less) information," it would be better to consider the different qualities of audio and video recordings as characteristics of distinctly different frames for data collection.

In this instance, the camera was so constrained that the separateness of the two was evident in viewing. With the camera unable to track the speakers and the video frequently decoupled from the audio, then joined together when Sergio once again began speaking, it required little effort to maintain an ethnographic distance from the scene. It was impossible not to notice the presence of the intervening camera. Thoughts about the way video configures data quite naturally spring to mind.

Within this setting, a question like what Sergio actually said seems weak and anticlimactic. He did verbally describe the fourteen options, but he also managed, pointed, was at ease or unease, sketched out, elaborated, queried, directed. Watching the replay is like walking through a funhouse room of mirrors. Sergio's image dominates; he is elongated, stretched, bowed, and all there is to see.

Others at the meeting are off to the side, like a chorus. Where was Fritz's frustration with the lack of support for the electromagnetic excitation option? (Recall, I was there. It's in my notes.) Who was listening? How involved was Hans? The video record has the character of a visual trick, antisocial, powerfully focusing your thoughts as well as your eye on the talking head and in this way distancing you, the viewer, from the true ambience of affairs, the more egalitarian confusion of voices.

Being there, I know that the scene was different. Sergio, although standing up front and in control of the flip charts, did not appear head and shoulders above all the others. I could identify, and did so in my notes, separate coalitions of twos and threes around and across the table, forming and dissolving. Out of the tête-à-tête would come a question that might lead the discussion off the agenda, and Sergio would then work hard to get it back on course. The expertise of different participants could be identified in these clusters by the nature of their questions and comments. A network rather than a chorus was the better metaphor, a network allowing occasional cross talk, as well as coherent communication from one node to another, but not always including Sergio.

Viewing the video, I note that there were moments when Sergio paused, tossed the marker from one hand to the other, tugged at his mustache, looked ahead through the flip chart, and then spoke, sensing an opening to regain control. On the other hand, being there at the meeting and able to track each speaker participating in the dialogue, I did not see Sergio's fidgeting or notice the unease that is so clearly evident in the video.

Video data are different from audio data are different from the leafs of notes I hastily scrawl at the same meeting. All require a framework of contextual knowledge in order to be read and decoded. Being present at the event contributes; so, too, does interviewing participants before and after. Then, too, one would expect that the more frequently one is on site, the richer one's basis for contextualization. Being an engineer—one of them in a sense—contributes, but even with all of this I remain a "fly on the wall," an alien of sorts, with my own preconceived notions of what is important to design process:

The attempt to [gain] a knowledge of reality, freed from all presuppositions, amounts to nothing more than a chaos of existential judgement (*Existenzialurteile*) based on innumerable particular perceptions. . . .

The reality of each particular perception always presents to us, if we examine it more closely, an infinite multitude of singular elements which do not allow themselves to be expressed in an exhaustive manner in perceptual judgment. The only thing that brings order to this chaos is the single fact that in each case, only a portion of singular reality is of interest and significance in our eyes because this single portion is in harmony with the notions of [what is of] cultural value with which we approach concrete reality.[15]

It is our expectations of the whole, the gestalt, that lends credence to the claim that there is dropout in the printed image. To see the blemish it is necessary to see the whole in anticipation of the missing. It is not enough to look at the minute pixels of light and dark that make up the image individually, even if all are within our field of view. So, too, when we look at design process using all forms of data—audio, video, direct observation—a framework is needed to see what is not there.

It is doubly important, when the focus is on technology, to attend to the framing. Technology, as it is commonly perceived as machinery or a set of rules, is something outside of us, out there. It draws our attention to itself as a thing apart, operating in rigidly determined ways, repetitive, and usually nonnegotiable in human encounter. This vision of technology does not serve if my intent is to understand design process: It disallows taking seriously the ambiguity and uncertainty of process recorded in my notes. This is what is not there in our usual images of technology; this is what I try to capture in my encounters and what is essential to understanding design as a social process.

Amxray

The facilities of Amxray Corporation preserve the character of the warehouse complex they once were. The firm purchased the abandoned buildings some ten years ago, then renovated them extensively. The interior was completely gutted, interior walls relocated, the remnants of electrical and plumbing networks brought up to code and thoroughly revamped to supply the high-technology needs of a laboratory as well as Amxray's production facilities. I enter the receptionist's area, sign in, pick up a badge, clip it to my jacket, and wait for an escort.

Today Michael, the project manager on the new cargo inspection system, comes down to the lobby, welcomes me, and takes me through the security doors into the engineering wing. We move down a long hall running the length of the main building, parallel to the now unused exterior railroad track. A series of bays open out onto the hall. Chest-high walls provide some privacy for the four or five individuals housed in as many cubicles within each bay. A steep, sloping glass roof rises overhead and meets the second floor. We are headed down to Purchasing at the end of the hall; it seems like a football field away.

Michael wants to ask Jerome about the promised delivery date on the detector crystals. A field agent of BG, the prime contractor on this job, is due this morning to review the status of the project, and Michael wants to be as candid as he can be.

Michael *Jerry, when are those crystals going to arrive? Have you heard anything new from SiGer?*

Jerry *Nothing new since I called last Friday. They claim they are going to meet the delivery date they promised in their last letter.*

Michael *Do you believe that? I'm a little uncomfortable with their request to change the specs. Did they buy what we wanted?*

Jerry *That presented no problem. They say that's a minor change in fabrication. I believe them. They've been true to their word in the past.*

Michael *So if I give BG SiGer's date, I'm as straight as I can be? No need to pad a bit? I don't want this coming back on us in two weeks time; and those crystals still haven't shown.*

Jerry *I'd go with that.*

We go upstairs to Michael's office. Much the same arrangement upstairs as down, only at this level we go by the bays of other project managers, of the firm's comptroller, its corporate lawyer, the marketing manager's office, and the president's suite.

As in many engineering firms, a dual hierarchy is set up to focus the energies of specialists—scientific, engineering, managerial, marketing—on the specific design task. Michael's team consists of individuals drawn from different organizational niches: Tony, from Mechanical Systems, has responsibility for the design of the detector support structure, the mechanical requirements of the x-ray beam collimator, and, for the purposes of the demonstration unit, the mechanism intended to carry the cargo container past the x-ray beam at a prescribed rate of speed. Jim, from Electronic Systems, has relatively more of his time committed to this project. He will, with three of his people, be responsible for the design of both the analogue-to-digital (A-to-D) processing hardware and all the software that transforms these data into an image on a video monitor. Arnie, the project scientist and a Ph.D. physicist, is responsible for a broad range of tasks including defining the power of the x-ray source required, fixing the best positioning for the precollimator, and, with Jim, figuring out the number of detectors, the rate of sampling, and the specifications of the data manipulation software needed to ensure an adequate image.

Frederick, a specialist in materials, knows all about the properties of the detector crystals, physically as well as electrically. He participates in discussions intensely now. Later, he will back off, once the crystals are in hand and have been proven "up to spec."

Michael *BG's field man is due in any minute. Got to go downstairs and check with Frederick before . . . Want to come along?*

This time our path leads to the back side of the plant, into the production facility, through the laboratory. Frederick's office, a window-less interior cubicle, shows little shelf space unoccupied. There are instruments and tools, bits and pieces of hardware strewn about on a desk that extends around the room. Shelves above hold large binders of specifications and codes.

Michael *Fred, how does that first shipment of detectors for the BG project look to you? Have you been able to get a reading?*

Fred *They look fine. There was only one that looked low on the response spec, but that may have been in my connection. I'd say we have a third of them in hand ready to prepare for insertion into their holders.*

Michael *Can we show Alex anything this morning? He's coming by.*

Fred *Bring him around. I've got to meet John on that Nevada project but you might try me. Best give me a call.*

Michael looks up. A paging voice asks him to call the receptionist. Rather than respond, he heads for the front entrance, knowing that Alexander has arrived. Along the way he stops at Jim's bay, spots his point man on electronics, and goes in to query him on the detector cabling.

Michael *Hey Jim, have you got all you need to connect up those first banks of detectors. It would be nice to be able to check them out this week, integrated with Mark's A-to-D stuff.*

Jim *We've got a problem there; you know that the J-tool is out in Nevada.*

Michael *I didn't know that. Isn't there another one around here?*

Jim *Not that I've been able to find. But Tony has a call in to their Connecticut branch. The tool only costs ninety-five bucks; we can afford that, can't we?*

Michael *Money is not the problem. Time is. We've got only four months left on our six-month budget. Why not make the damn thing downstairs? I've got to go pick up Alex. We'll stop back.*

Michael greets Alexander and takes him up to his office. They review a printout of the status of all purchase orders and delivery dates, noting the redlined items that may present a problem. Michael adds a red marking to cabling connectors. Then back on the road again to purchasing. "Yes, the detectors look good. We trust them for the balance of the delivery." Over to Tony in Mechanical Systems, who queries Michael.

Tony   *Did Jim tell you about the problems we're having getting hold of a J-tool? If worse comes to worst, we might have to try to make one in the shop.*

Michael   *I filled Alex in on that one. It would be a new one on me if this two-bit tool makes us slip a week. You've got a call in to Bridgeport?*

Then on over to Frederick's lab, to view the crystals; then back to Jim's to get a fuller rundown on the cabling scheme, the interfacing with the A-to-D unit.

In the midst of these fragments of conversation, of questions and responses about the design and development of an x-ray inspection machine for cargo containers, I try to form a global vision of the project and more specifically attempt to fix the current stage in the design process as Michael makes the rounds of the firm. Within these eddies of interaction I see and hear reference made to bits and pieces of hardware and software, tooling and suppliers, specifications and deadlines. Micromanagement: Is this the design process? How do these bits and pieces constitute a whole? Is this the way in which technology is defined and shaped to useful purpose? Is this discourse designing?

I consider backing off from observing the day-to-day hassles that at first seem so far removed from what I had imagined, and what the textbooks describe, as the design process. I'm tempted to renew my study from a position more distant. I could, for example, learn about Michael's work and role as project manager from formal documents, about all of the other actors from the firm's organization chart, about the technology from the proposal that defined the task, about the chronology of events from the schedule that lays out the order of work on the different subsystems—all hard forms of archival data. I could analyze and explicate the content of these documents, important enough to the process to have been made formal, rather than try to read the scraps of sketches, annotations on computer-generated lists, and the like that rarely survive the day of their use. I could seek further elaboration and explanation through scheduled interviews with Michael and whoever else has time to spare—get them talking *about* the design rather than listening to their discourse *within* the design. And when it was all over, I could view the prototype inspection machine in operation; and they would explain to me why it was so and how it came to be that way. In this way the cut between subject and object would appear to be clean; there would be evidence, records on paper, and other identifiable data bits to accumulate and cite. After compiling and sorting things out, I could construct and analyze a chronology of causal relationships and critical decisions in fabricating a story about engineering design.

I remain uneasy in the light of what I have seen. I am leery of the neatness of this alternate approach; I judge it inadequate to the task I have set, which is to understand design process in its full complexity. Furthermore, I am intrigued by the day-to-day hassles. There is an energized air of uncertainty and excitement in even the most casual meeting I witness. Again, a flurry of questions races through my mind: Where, and what, is this sacred J-tool? What does all of this talk have to do with the definition of the hardware or artifact? Where is the design? What is the significance of what I do observe?

I decide to stick with it. I first set myself the task of setting straight in my own mind the x-ray machine they are designing and how its subsystems are meant to function together as a whole. I reflect on the hard object, the focus of their conversations, sketches, phone calls, and computer printouts. The physically most massive piece is the tall, boxlike structure that houses the x-ray source. The track for the cargo containers runs in front of it, and on the other side stands the column of crystal detectors that sense the spatial variation of intensity of the radiation that has made it through the cargo container. The computer that does the data collection, conversion, and image processing sits some distance away.

I muse about imaging. This machine is quite sophisticated. An x-ray image even of a suitcase on its way through U.S. Customs often surprises. Shaving equipment and hair dryers don't look like themselves, nor does the traveler usually have the opportunity to see the ways in which this kind of image can be further massaged and remade by a competent operator at the controls. It takes time to learn how to manipulate all there is to manipulate in order to extract out of the raw recorded and stored bits of data a representation that suggests a bomb, a package of a powdery drug, or the image of other contraband. With "contrast stretching," "edge enhancement," "zoom," and, in Amxray's most sophisticated machinery, color coding, the variations on an image appear unlimited. A truly kaleidoscopic set of representations of things hidden to the naked eye can be constructed with the crafty turning of a few knobs. Watching the project scientist, perhaps the most experienced operator at Amxray, shape and reshape an image displayed on a monitor prompts me to ask what I am really seeing. What is really there?

As participant/observer of the design process, I am also an instrument. I must learn how to perceive in the ever-changing display before me the meaning of these variegated events and their significance to designing. I must learn to discriminate important technical statements from other kinds of statements. What is foreground, and what is "background noise"? But I must also become attuned to context and test the latter, recast the noise to see if it might have significance from some other, broader perspective.

Engineering Observations

By themselves alone, these reports of designing drawn from ordinary life within Solaray, Photoquik, and Amxray seem of little substance if our intent is to construct a coherent general description of the engineering design process. If we limit our attention to the technical content of these narratives, expecting to formulate some significant insights, we are bound to be frustrated, left with but three sketchy descriptions of the design of a photovoltaic array in Saudi Arabia, a dropout problem, and the readiness of the pieces of an x-ray inspection machine for large cargo containers. A good bit of uncertainty permeates these discussions. The talk we hear about the range of possible sizes of a photovoltaic array, the whereabouts of a J-tool, or about shifting deadlines on hardware mock-up and testing is highly tentative. Statements appear as mere pointers toward the specification of characteristics of design and are hardly definitive.

This lack of definition derives, in part, from considering these reports out of context. That is why I have chosen the role of participant/observer, intending to be around frequently enough and over a span of time sufficient to allow me to fill in the background, to fix the references, and to complete the statements left hanging, just as learning to read an x-ray of a fractured femur requires the radiologist to confront a sequence of "before and after" images. But it also derives, in large measure, from our traditional view of technology—a deficient view in that it rules out the possibility that the fragments of muddling through and hassling about that I have observed are important and necessary ingredients of the design process.

It should be clear by now that I have trouble with our traditional ways of seeing technology. When I go into the firm, seeing in this traditional way, I have difficulty reconciling the maze in which I find myself with what I have been schooled to see. I have the feeling, in making the rounds with Michael or tracking the discussion of redesign options with Sergio, that I am going up a down staircase in an Escher engraving, rather than neatly traversing an engineering textbook's block diagram of the design process. The traditional way of thinking and speaking about technology is simply not a sufficiently rich basis for the study of designing. A less neat vision of technology is required.

It is understandable that we encapsulate an x-ray inspection machine in our thinking as hardware, that we draw an envelope around it as if it were on display, sitting in a showroom or installed at an airport or a science museum. After all, technique is usually made known to us through some sort of physical presence in space and time; that is, it arrives in a box, gets unpacked, is set up and then turned on. Even if it is a complex system, such as a telephone, or a set of procedures, such as the rules for setting automobile insurance rates or for governing financial transactions at the local bank, there is at least a coded form to fill out or some artifact to grab onto, climb into, or push the buttons of.

Setting technology apart seems necessary if we are to make sense of it. We learn by categorizing as different all the bits and pieces of our experience.[16] But in this business of classifying, it may be that the glasses we wear, the filters we set before us, or the sieves we use in collecting, sorting, dividing, and piling up our experiences and observations are not the right tools. Our taxononomy is not as intelligent or as meaningful as it might be. In particular, when we look at the contemporary world and see technology, we often oversimplify and split the world in two: C. P. Snow's two cultures; university programs in technology and society; engineers' hard and soft (all of which is not to say that this kind of categorization is without significance or utility).

For sure, the knife has served us well in science. A good bit of the Lavoisian chemical revolution consisted of sorting out stuff, untangling substances, and naming them as elements, distinguishing these from the apparently more complex and composed mixtures that swirled about the Enlightenment chemist's laboratory in gaseous, liquid, or solid form. The edge of the knife was honed according to Lavoisier's criterion: The elementary could not be reduced or broken down further through chemical means. With this set of elements in hand, we have a way of seeing, organizing, and understanding the myriad of materials around and within us, a way that satisfies our need for order. It also proves a basis for the production of more myriads of new toxic, exotic, and useful materials. In all of this it is the elements that we hold on to and that give, at least to chemists, a sense of control. We claim that compounds are made up of elements; the former can be broken down into and reformed from the latter. At the same time we claim that the elements somehow retain their identity when compounded. It is a bit like claiming to have your cake after you have eaten it.

A constancy through time acts here, and not just a conservation of mass. Knowing that we can get the elements back out of the compound, and exactly the same amount of those elements and no others, we continue to "see" the elements in the compound. And this despite the fact that the properties of the compound are manifestly different from those of the elements. In short, transmutation is not allowed.[17]

There are very good reasons why we do not see the synthesis of elements into compounds as irreversible. Without this constancy through time, we could hardly have a science of chemistry. Indeed such a "one-way" world would be totally inconsistent with modern society's faith in some kind of eternity of physical reality. No, we must be able to take compounds apart by chemical means to complete the process, no matter how arduous a task it might be—indeed, even if it is not possible with contemporary means.

This faith in the continuity of experience is given witness throughout history in attempts to seek and construct a deeper structure that might explain how the chemical elements interact and form the compounds they do. Hence theories about intercorpuscular forces, or about "affinities," or eventually "valence." At this level, below the first cut of the knife, we find we can talk about the compound as being a kind of summation of the elements according to some prescribed rules that govern their play.

So, too, we see technology as an element in society that combines with others—political, economic, bureaucratic, ideological—and changes the compound that is culture. We insist on keeping its identity intact amid the complexity of culture. We speak as though we can always do the reduction, point, and say, "There is the technique—there, out of that box—that has caused all the problems (or led us to these new heights)."

We further complement this with a scenario about how the elements technology and society interact: A technology (telephone, incandescent lamp, horseless carriage, nuclear power, etc.) hits the market. It has impacts on society. The impacts can be qualitative and ideological as well as economic. Society, in turn, is powerfully enabled to do new things, to get things done more quickly, or to reach further, dig deeper, and so on. This reorients market forces, may disorient peoples' lives, and stimulates a demand for new technique varying in kind or degree. And so on.

Yet it is clear that the process is more complex than what this particular and common reification of technology implies. The reaction is not reversible; we cannot withdraw the technique and return to try to do it better. Still we persist in seeing technology separated.

I want to break away from this "technology as element" vision of technology in culture. I will adopt a more alchemical perspective that allows for transmutation and recognizes the irreversibility of history. I want to frame my study so that technology is more integral, more ideational, more fully social than is allowed by a framing that sees it only as a material element of culture.

To do this within the firm, I must become attuned to the different levels of meaning in a statement, even if superficially it appears as a statement of technical fact alone. I must begin to search out connections among apparently straightforward technical statements and the interests of the different participants, ask questions like: Why are they making this particular topic a priority? What resources do they call upon to justify their claims and proposals? How do they organize with others to effect their plans? This will require some sort of image enhancement since we are so accustomed to focusing on the technical threads alone.

Let's start by noting that, while participants in all three accounts talk about *hardware*—batteries, photovoltaic arrays, control subsystems, J-tools, crystals, rollers, baffles—these things are not so solid and well defined as the word suggests. They are continually referred to, brought into question, explained, elaborated, even denigrated. There is considerable uncertainty and tentativeness about these things; the talk is about how a photovoltaic array of this or that size *might* work, how much it *might* cost, when it *might* be delivered, which one *should* be chosen. Technology defined lies off somewhere in the future.

Time has an especially urgent status within these affairs. One senses that there is not going to be enough of it. Each of the three groups feels the pressures of time. The future is not to be approached leisurely. There are deadlines to be met, problems to be resolved, connections to be completed, reports to be written, deliveries anticipated, delays to be allowed for. Where is Beth?

Participants seem to be on some sort of preprogrammed trajectory toward closure. Michael, Brad, and Sergio, as project leaders, strive to establish and maintain coherence, to build networks of things, people, and interests that will hold together over time, or at least until the next meeting. At Photoquik, Sergio struggles to narrow the options; at Solaray, Brad strives to pull together all the pieces of the proposal; at Amxray, Michael makes the rounds, constantly reconstructing the status of the design of the cargo inspection system. Each has a vision of what the final state of affairs should be, which is to say a product, or at least a prototype functioning in the world, outside the confines of the sub-culture. That vision is only partly encoded in a set of performance specifications and drawings contained in a contract. The process of designing is a process of bringing to life and adding flesh to these dry bones. Participants know where they are headed. They all anticipate the launching of the product, the day "their baby" goes "out the door." Delivery to a customer or a contractor will mark closure. Whether or not they will get there on time, under budget, or even make a successful delivery at all is, during the process, an open question. The traditional view of technology would say otherwise: Instrumental technique alone should suffice to go from a set of specifications to a functioning product.

The apparent incoherence and uncertainty of the process I observe derives in large measure from the differing interests and viewpoints of different parties to the design. This is not simply due to the lack of detailed specifications and analyses of how the technology will, or will not, work, nor can it be written off as a methodological artifact, although initially, no doubt, my unfamiliarity with context played some role in limiting my perceptions.

The resolution of this uncertainty is very much contingent on social connections and networks, on the relationships of participants in design to each other and to outsiders. When is *Ed* going to develop a better estimate of array size? How is *Sergio* going to get his *group* to agree among its members to two or three options? Is *SiGer* going to be able to live with the new crystal spec? The three narratives call into play a spectrum of interested parties both within and without the firm.

Their discourses are multivalued; they reveal different meanings at different levels. For Michael to consider making a J-tool in-house is more than an instrumental trade-off of expedience against cost. It is an issue of Michael's and his team's power and authority within the firm, a potential test of their ability to command the firm's scarce resources in competition with others (that other group in Nevada). At still another level it is a display of the group's need to maintain command and control over the uncertainty of the design as it unfolds. At still another level, the way the question is raised and promulgated among members of the design team—as potentially the straw that will break the camel's back—points to its use by participants in maintaining the momentum of the project in the face of all the other things they have to do. It is a rallying cry.

Similarly, at Sergio's meeting to narrow the options at Photoquik, lurking behind an apparently straightforward question about the power requirements of one of the fancier options he offers as a possible fix for Atlas product's dropout problem lies the reluctance of the Atlas product development team to consider any unproven option, especially if it will require them to redo some of their work. The individual who raised the question was the Atlas program's liaison to Sergio's effort. The mild-mannered way he posed the question can be read as a disinterested scientific query, but it can also be seen as an attempt to make alliances with others who, for their own reasons, have serious questions about committing resources to an approach that is too novel. Sergio, in turn, must be careful in his response not to cut off this challenge. (That he favors this option is clear from the way he has presented the set.) He needs to tap the resources Atlas controls to justify his efforts; they are scheduled to provide him with the machinery and a technician to test out his options. And he must work hard to build a group identity, work against the fragmenting of participants into separate subgroups interested in just one of the options.

Even the talk at Solaray about the weather in Saudi Arabia and Phoenix is multivalued. It at first appears as little more than banter. As such it has a certain social value; it is a way of keeping the exchange well tempered and open. But banter about high winds can have technical import: The intensity of windstorms in Saudi Arabia can be a major factor in the design of a photovoltaic system. Wind loads are a serious consideration in the design of large structures supporting large flat surfaces, like photovoltaic panels, and windstorms can leave a layer of fine sand over the module's surface, degrading its performance. How many modules are required then? What is a sufficient structure to support the array? This light banter can lead to more serious talk at some future gathering.

At another level: Ought one to talk about the weather in the production of the proposal? If Brad and Solaray don't discuss the weather, they risk being seen as incompetent, their proposal underrated. Yet to address this item in more than a superficial way will require someone to dig out the frequency and intensity of windstorms and to check out estimates of degradation in array output due to dust accumulation; all of this will require time. Weather banter, then, can be a prelude to "serious" technical discourse. Just what participants think of as a robust design, and what as a marginal design, is lurking behind words spoken in jest. Through these apparently trivial exchanges, a framework for decisions is given shape and the character of the work of the design team is revealed.

In all of this, facts, values, and individual interests are intertwined. Ought one to make a J-tool in house? Use a special module price? Is that a conservative estimate? Can we take a chance and risk resources on an innovative approach using an electromagnetic excitation device or should we stick with the traditional, albeit clumsy, fix? These questions and the responses constructed by participants depend as much upon their individual beliefs about what is best, upon their shared trust in what may be possible, as upon fact, instrumental analyses of the way things will work, or quantitative estimates of their costs and benefits. These estimates in themselves are ultimately rooted in participants' faith and trust in some sort of constancy of circumstances trailing off into the future.

We begin, now, to see the design process in all of its dimensions. It is only from this broader perspective, one that encourages an understanding of technology as object and process mixed, that these vignettes make sense. At first reading, their recording is messy and undistinguished. Too instrumental and one-dimensional an interpretation leaves one with too crude an image. To render meaning, I must admit the possibility that design is complex, social, and uncertain. I need a new framing of technology that sees what is not seen in order to produce a thick description. As anthropologists must attend to the mundane in the construction of social fact, must see in the material culture of objects of use and ritual an embodiment of values as well as practical efficiency and artful creativity, and must recognize as meaningful what the visitor too often dismisses as irrational, mystical, or unscientific, so too I search out in these seemingly opaque events the necessary ingredients of engineering design.

This is the framing I choose. I want to uncover meaning in the (sometimes irrational) practices and events I observe. I want to make life within the firm understandable (rational at another level and from another point of view). I want to construct a more useful story about the determinants of the form and function of technique.

I have described some problems in method—the problem of access to events, the problem of framing the events I do witness, and the problem of generalizing on the basis of discrete and limited observation.

In some respects these problems are characteristic of the natural sciences. They take on an immediacy when the phenomenon is social, when the controlled laboratory experiment is not possible, and when the native studied is like you the observer.

These problems of method have implications for analysis and presentation. I will rely on narratives of events as I observed them to convey the story and provide evidence. Narratives are not transcriptions of recordings but, as I have said earlier, productions, my best attempts at decoding design process. These accounts of process observed within the three firms, in turn, taken as a whole, and through contrast and comparison, become the basis of analysis on a more abstract level. That is the gist of the method I will follow.

I coin one jargony phrase in this book because it seems to stand up throughout my observations. I will refer again and again to the object world. That is where I will start, back with the object, but now with the intent to break free and see the process of design anew.

3

The Object

Energy Flow

Beth sits at the lone computer terminal left in the room. In fact, it, the table it rests upon, and a chair are the only furnishings remaining in what was, up until this week, her and Tom's office. Solaray is rearranging its work space—not for the first time or the last—and Beth's group is moving from this now near-empty bay adjacent to Cell Testing to another building. Because the terminal in her new office has yet to be connected to the firm's mainframe computer, she must put up with the inconvenience of this barren space in order to pursue her work. She has put off for too long the development of a computer model of Solaray's photovoltaic-powered desalination plant installed last year in Saudi Arabia.

The plant was intended to demonstrate the potential for photovoltaics to play a significant role in the conversion of salt to potable water. It was a prototype design, the first of its kind for Solaray, and its performance has yet to be evaluated. Beth's computer model, together with data collected at the site, would provide the basis for evaluation of this plant and for the design of the next one.

The plant has had its rough moments. At the beginning of operations, a little over a year ago, Tom had spent the month after start-up nursing the system along. Now he talks about that month with gusto, relishing the quirks of performance and the challenge of communicating with both the local "technicians" and the home office, dwelling particularly on the moments of creative improvisation that finally led him and his teammates to a remedy of the problem. This in truth was not as easy a fix as the company lore suggests. Out in the field, a long way from home and all the usual resources—colleagues with know-how and expertise, instruments, and documents—pinning down the cause of the faulty performance proved a serious challenge. The failure of the photovoltaic array to produce as much power as it was designed to produce could have been caused by a variety of conditions; faulty workmanship on site was more likely than a flaw inherent in the design.

As it turned out, the problem lay in the design. The drop-off in power production due to the heating of the cells above ambient temperature—the hotter the module, the less power produced—had been underestimated. In the Saudi midday this can be considerable. It took another month to effect a fix; an additional module was added to each series string. Now, a year later, the plant had produced a good bit of potable water. The on-site instrumentation had, for its part, produced a year's worth of raw data about current flows out of the array, into and out of the bank of batteries and to the pumps. Data about voltages at key points in the electrical circuit, about temperatures of selected photovoltaic modules, and about rates of flow of the treated water had also been recorded, all at five-minute intervals. Still other special instruments had measured the intensity of the sun's radiation impinging upon the array.

It is Beth's job, using these data and the underlying form describing the behavior of photovoltaic systems, to construct a computer model that tracks the flow of energy emanating from the sun and describes how it has been transformed by the photovoltaic array, stored in and extracted from the batteries, and then expended in powering the pumps that produced clean water.

Beth must describe where all the sun's radiant energy falling on the array goes. In particular, she must account for all possible ways in which energy might be diverted or wasted and develop estimates of these "losses." She knows that a photovoltaic module nominally converts only 10 percent of the sun's energy to electricity; the balance is either reflected or transformed into heat in the conversion process. Batteries are not 100 percent efficient; nominally 80–90 percent of the energy stored is available to be drawn back out and put to use. The pumps also waste some energy in operation, and even the electrical wires dissipate some small percentage in conducting power from one subsystem to another.

Ultimately every bit of energy must be accounted for. What is lost is attributed to an inefficiency, and these inefficiencies must be in accord with the accepted, historical performance of the individual components or subsystems. Like a material substance—water flowing, for example—energy doesn't just vanish, but it can slip away from your grasp. The principle of conservation of energy reigns supreme within the world of the engineer.

While this principle provides the underlying form girding the model, the crafting of a computer representation of the behavior of the desalination plant in the field is not a paint-by-numbers activity. While she does have nominal values for the efficiencies of the array, the batteries, and the pumps, Beth wants to construct a more detailed and accurate account.

In this she has two objectives. First, she wants to verify the quality of the data produced by the on-site instrumentation. These data alone might be used to ascertain the worth of using photovoltaics as an energy source for the desalination of water at a remote site, but only if they are reliable. Second, she wants to construct a computer model that might be used with confidence in the design of other photovoltaic systems for remote sites, in particular for systems that, while also intended for desert climates, are significantly different from the current prototype. Both of these objectives require her to construct appropriately detailed representations of the behavior of each and every subsystem for computer processing.

More than nominal values for efficiencies, her model must be able to account for vagaries as well as predictable changes in the functioning of a photovoltaic module, a battery, a pump, or a control unit as things change from minute to minute and day to day. Relating the measured field data back to the computer model predictions at five-minute intervals gives her a way to validate the data. At the same time, if Tom wants to know how the system would perform with one less battery in each series bank of batteries, Beth can simply alter this number within her computer model and provide him with an immediate answer. The incestuous character of this model-making process—the model designed to verify the field data; the data, in turn, providing a reference for the model—is noteworthy.[1]

Beth switches on the terminal, logs on to the mainframe computer, and calls up the data files. She scans through these one by one but finds that there are gaps. Why are there gaps? Are the data simply at another place in the file, or were they not read off the tape and into the computer, or is the lack of a record due to no tape being made in the first place, perhaps reflecting a failure in the instrumentation? She opens her copy of the site logbook. The first ten pages are filled with Tom's scrawl, then with copies of the forms filled out daily by a technician. Each sheet shows selected bits of data, redundant with what the data collection system much more frequently recorded, intended to help the system operator keep track of performance in real time. Beth rifles through these.

She locates the days of missing data. On three of these days the forms note that the system was turned off, "down for maintenance." On two other days they mention a problem with the data recording computer. The remaining two days appear to be normal. She can easily accommodate the days the system was down, but she decides to try to extrapolate over the two days where the lack of data was due to failure of the data collection computer.

She gets a call from Brad: Drop everything and come help work up an estimate of the costs of a photovoltaic system intended to protect an oil pipeline, running miles across and under the Saudi desert, from corrosion. Another proposal, this time on cathodic protection, must be sent off on short notice.

One thing leads to another, and a week goes by before she returns to her evaluation of the performance of the desalination plant. She has to refamiliarize herself with the records of data, where they are, and how she was going to manipulate all of this information.

Completing this inventory process, she addresses the more intellectually challenging task of constructing models of the behavior of the system and its ingredients. She turns to writing a computer program, a coded sequence of manipulations of the field data in accord with the principle of conservation of energy that will allow her to construct a full picture of performance in terms of the energy flow from the sun onto the array, to the pumps, in and out of the batteries, and so on.

She has an old program on file, one she had begun to write before the plant had been turned on. She had intended to have this ready to evaluate the performance of the desalination plant week by week in "real time," but, with the problems of start-up, Tom had suggested she put this on hold.

Now she tries to run the program but discovers that the format of the data as recorded on tape, in the field, is not compatible with what she had expected and programmed for. This is a minor problem, easily solved by changing a few format statements in her program.

The phone rings. It is Brad. "Do you have the test results on the voltage regulators run in the 'goat yard'?" A distributor in Arizona has gotten a call from one of his major customers wanting to make sure that his batteries won't be overcharged. Another fire to put out.

It takes Beth but an hour to respond. Tom had set up the tests; she had done the data analysis and summarized the results. It's simply a matter of printing out a copy of her report and handing it over to Brad. After lunch she returns to her program, and by the end of the day she has a crude version running.

The main code links together several blocks or subprograms. Each block describes the performance of one of the subsystems, which, when taken together with all the other blocks, constitutes a representation of the entire desalination plant. The block that models the photovoltaic array requires, as "input," average numerical values over some time interval (which Beth had chosen as five minutes) for the ambient temperature, the direct and diffuse amounts of solar radiation, and the voltage "seen" by the array. The core of this block, the relationship between current and voltage and how this relationship depends upon these input values, yields an estimate of the average current the array produces during the five-minute interval. A further computation gives the energy produced by the array.

Another block models the behavior of the batteries. It describes how much energy is lost in passing in and then out at some later time, and how the battery voltage depends upon the rate of charge or discharge. A third block models the workings of the pumps and the desalination machinery as they filter and process the water.

Each block requires information from the other blocks in order to play its part in simulating the operation of the plant. In particular, the estimate of the current produced by the array depends upon the voltage the array "sees" when it looks at the batteries. The voltage "shown" by the batteries depends upon their "state of charge"—whether they are near full charge or near empty—whether they are powering the pumps and, if so, how much current the pumps require. (This means that Beth's model must account for the cumulative effect of charge and discharge over time.) With the blocks tied together, the program will compute the values of all the internal measures of performance at each instant given just a few external values—namely, the solar intensity, the ambient temperature, and the flow rate of the treated water—all of which were measured in the field.

Beth will use the other measures taken at key points within the system—the current powering the pumps, the voltage at the batteries, and the current flowing into or out of the batteries—to check her computer models of each subsystem.

As soon as she has constructed and tied together all of her program blocks, she selects a data file from the early months of operation and executes the computer code. Not unexpectedly, the numbers appearing on the monitor screen don't look right; when she adds the measured current flow into the batteries to the current flow into the pumps, at some arbitrarily chosen time, to obtain a field measurement of the current produced by the array, and then compares this number with the value calculated by the subprogram modeling the array, the results are inconsistent. The array output as obtained from the field measurements looks excessive. The model for the photovoltaic array predicts less current, although Beth has used values for the ambient temperature and the solar intensity measured in the field at the same time.

What is the source of this discrepancy? Is it a fault in the model or a fault in the field measurement and recording hardware? Beth cannot believe that the array could have worked as efficiently as the measured current values indicate, especially after hearing Tom's tales about the problems he had encountered at start-up. Could the instrument that measured the intensity of solar radiation be the source of the difference? Perhaps it was dirty, giving too low an estimate of the energy available from the sun, even though the on-site technician had been instructed to clean the cover glass each morning?

That night Beth sleeps lightly. When the baby wakes her at 2 A.M., she has trouble getting back to sleep. She arrives early at work the next morning and heads for the computer terminal. But where is it? Yesterday's workplace is today truly an empty office; her terminal has been moved to her new office. There she finds her colleagues Tom and Ed gearing up for the day. She powers up the monitor but is frustrated when she attempts to log on to the mainframe. The machine doesn't respond. Ed explains: The cabling connecting their office back to the mainframe computer has not been fully installed. Furthermore, the "hard-wiring" won't be completed and the connection made for another week.

Beth locates a functioning terminal in an office adjacent to her old office. Bill is out today, and Sarah says she is sure he won't mind if Beth uses his office.

She returns to her program and surveys the field data, looking for periods when the measured current flow into or out of the batteries is zero. At these times, all of the current flowing out of the photovoltaic array goes directly to the pumps; the *measured* current to the pumps should then match the computer model's *prediction* of the current out of the array. It doesn't. Instead, the computer model predicts an array current less than that measured going to the pumps. If the array model is taken as correct, the batteries should be discharging rather than showing zero current.

She muses: "Was the instrumentation faulty? Was there a bias in the instrument measuring the battery current, or was it the field measurements of solar intensity that were in error? A bit of an offset would account for the problem." She tries a few other days' data, looking for periods when the battery current was zero, when the array output should be equal to load as measured by the current to the pumps. She finds more mystery: Only in the first few months of operation does the discrepancy appear; after the third month of operation the measured current flows look consistent with her computer model of array performance.

She recalls seeing an entry about data collection in the log book. She finds the records from June, the third month of operation, and goes through them page by page, day by day. Nothing about fixing or refurbishing the instruments appears in the notes. Tom should know more about this; he was there in June.

Tom *Oh, yeah. We put in a new ammeter. We noted that the original was giving us some screwy readings after we added the temperature sensors.*

Beth *What day was that? What was the date?*

Tom *I don't recall. It should be in the log.*

Beth *I don't see. Here's something. Is this it? "Data collection off. Array cleaned."*

Tom *That's it. We cleaned the array when we added the sensors, and that same day we changed the meter. What's the date?*

Beth *The 26th.*

Tom *That sounds right.*

Beth looks for periods during the days prior to, and just after, June 26 when the measured battery current was recorded as zero. She finds some five-minute intervals on the 27th and some on the 23rd. On the day after the new meter was installed, an estimate of the array current obtained from her computer model compares favorably with the field measurement of the current driving the pumps. On the 23rd, the comparison is not good. She conjectures that the ammeter measuring battery current was faulty, giving a reading of zero when the batteries were in fact discharging.

To further substantiate her thinking, she adjusts a coefficient in her program, replicating the effect of the faulty meter, then executes her program for a full week of data recorded prior to June 26. With this change, the system performance now looks reasonable. She runs some days randomly chosen from the first three months prior to the day of the instrument change. In all of these "runs" the energy balance looks correct; the array efficiency appears in all instances to vary slightly about a value of 8.5 percent, as it should. Some minor variations can be easily accounted for by variations in temperature of the array, perhaps a dust storm.

Over the hump, she turns to writing more computer code and to fix the format of her graphs so that she will be able to present a succinct and artful account of the performance of Solaray's first photovoltaic-powered desalination plant to her colleagues and ultimately to corporate headquarters. She is on a roll, at least until the next glitch appears or the phone rings. She has developed a computer model of how the system and others like it work. The data, the record of performance of the desalination system in the field, both guide her in the construction of the model, validating the model's design, and provide the means with which she can evaluate the performance of future desalination, or indeed other kinds of photovoltaic systems. She has reconciled her energy accounts—field data and flow computer analysis conform.

The Photovoltaic Object World

The intensity and specialty of Beth's experience in accounting for the performance of the photovoltaic system invites categorization as work within a world apart from the everyday arena you, I, or Beth for that matter, inhabit. To underscore its uniqueness I give it its own name: I will use the phrase *object world* to designate the domain of thought, action, and artifact within which participants in engineering design, like Beth, move and live when working on any specific aspect, instrumental part, subsystem, or subfunction of the whole.[2]

The distinctiveness of this world apart is marked by the object *as hardware* at its core. Here, and in all that follows, I will mark out the object in the form of realized, functioning hardware, *as an artifact*. Artifacts are only part of object worlds. Photovoltaic cells, their assemblage in a module, other more ordinary electronic devices such as fuses and blocking diodes, the solid-state components of a battery charge controller, the batteries themselves, and the instruments employed to measure currents and voltages cast Beth's world apart as an object world within the broad domain of electrical engineering. Batteries, we note, are also chemistry.

Beth's world includes mechanical artifacts and apparatuses as well. She must come to know the pumps that process the water in terms of their performance characteristics. Only then can she relate the amount of potable water produced to the amount of electrical energy supplied. And there are solenoid valves, mechanical switches, and other electromechanical devices she must contend with in her object world.

But artifacts alone do not an object world make. To Beth the object *photovoltaic module* is as much the symbolic, mathematical relationships describing how the current produced by the module depends upon the sun's intensity, the module's temperature, and the battery voltage as it is the artifactual, physical panel in itself. The "I-V" (current versus voltage curve) as a whole can be read as a module's signature, an image Beth can sketch out from memory in a few seconds.

Other images define other objects in her world. The sun, on a cloudless day, becomes a single smooth hump on a plot of solar insolation over the course of the day. The batteries are characterized by a plot of voltage versus state-of-charge, with the vagaries of their performance indicated by different curves for different rates of charge and discharge. There is no chemistry here other than as alchemical reference when performance seems to stray beyond the curves. And Beth produces her own images in the construction of her model—graphs showing the way array energy output changes with time, the amount of water delivered each hour over the course of a week, and the swings in battery voltage as the object-as-system goes through its paces. These images are all ingredients of her object world.

The furniture of Beth's world also includes all of the techniques and methods she brings to bear in the construction of her computer representation of the facility. The computer language she employs, the handbooks and reference books that store all the known data that characterize the units that make up her blocks, even the behavior of the computer at the systems level, all influence her construction of a model of the desalination plant.

Other items constituting her world are more a matter of this particular installation. The field logbook, while artifactual, is too cryptic to admit a solo reading. Her own knowledge of the subtleties of behavior of ammeters and machinery for electronic data collection and processing, her interpretations of drawings and sketches of the plant layout, her expectations about the behavior of the on-site technician, and her talk with Tom aid in and are prerequisite to extracting meaning from the log. All this also helps constitute Beth's world of the workings of a photovoltaic system.

In addition to hardware, techniques and methods, empirical and personal knowledge, indeed preeminent within Beth's world-apart, is the principle of energy conservation and all of its paradigmatic apparatus. This is what provides the underlying form and is the basis of her work. This is what is encoded in each block of her computer program and the basis for the quantitative account of the system's performance.

Working within her object world, Beth goes about making up scenarios about artifacts and principles, physical concepts and variables and how they relate; in her photovoltaic object world, the principle of the conservation of energy provides the basic plot structure. In this, she is like all the participants in design, whether they are mechanical engineers, marketing managers, technicians, or project leaders toiling within their respective domains.

Beth's scenario takes a different form depending upon its audience. If the audience is the object itself, or a surrogate—a computer model, a scale model, a prototype—the form of the description is severely constrained and fixed by the rules of syntax and vocabulary of the object. For example, a computer model is written in a computer language, and a prototype allows only certain direct measurements of performance variables. If the audience is the author alone, then shorthand notes and sketches, a page or two of mathematical expressions and relationships and some marginal calculations, will serve to construct and sustain a manageable representation of the system. If the audience is a colleague on the project, a more discursive and complete account is required.

The account, whatever the form, is cast in terms of a metaphor—that of flow. Flow is particularly appropriate for describing a system in continuous operation in which energy is conserved. Through this metaphor, which contains and suggests more than the principle of conservation of energy seen in symbolic form, she renders meaning to the shorthand sketches and equations she has scratched out on a few sheets of paper in formulating a model of the photovoltaic system. It is the backbone of her computer model. It is the root metaphor for elaboration and conjecture in responding to Tom, Brad, and others on her way to the construction of an adequate representation of the performance of the desalination plant. Through this metaphor she appropriates the system and gains control over its different components and their behavior in conjunction, one tied to another.

In the course of work within object worlds, these scenarios are continually modified,[3] told, and retold. One version is tested against another. They are often given expression in actual bits and pieces of prototypical hardware or, as in Beth's case, reduced and translated into a recipe to be played out on the computer. Frequently they are fashioned for exchange with others, sometimes informally with a crude sketch, less frequently given a full formal exposition, although their telling and hearing is essential to the process of design.

Compressed Air

Beth's model making is part of an ongoing process in the design of photovoltaic-powered desalination systems. It is an example of successful story making within an object world. It is also a story *about* success, the success of a design concept—using a photovoltaic array to power a desalination plant. I turn now to another example of object-world work. Don at Photoquik is grappling with a fresh concept for the solution of the dropout problem. Here, while the account Don constructs is worthy, his conclusion is negative: The air knife won't solve the problem.

Don had come late to the project. Sergio's mandate to come up with a fix for the dropout problem on the new line of photoprint processors had been established a month before he arrived. But Don was not unfamiliar with the product; he had been with Photoquik ten years, the last two in West Germany, and he had participated before in this kind of patchwork design activity. He was eager to come "up to speed" and begin work. He looked forward to the challenge of helping bring order to what appeared at first encounter as a rich mixture of technical uncertainty and bureaucratic confusion. Sergio, to acclimate Don to the project, asked him to evaluate the air knife—one of the fourteen options discussed at an earlier group meeting.

The air knife seemed like a strong contender for solving the problem, although Don had heard of the objections the people over in Chemical Processing had raised to its adoption. Still, at this preliminary stage of the design, it was worth spending some time exploring its feasibility.

Sergio had given him some rough numbers defining what force levels would be required to hold the paper down on the bed. So Don pulled out his old fluids textbook and estimated what air flow at what angle would produce a force of this magnitude. A bit more analysis, linking the flow rate and an estimate of the duration of the force to the static conditions of a plenum, gave him a rough estimate of the required volume of the air tank. He also figured the size of a compressor that would be needed to compress the air to the desired static pressure in the plenum. Don took his result to Sergio:

Don   *Here is the air knife, Sergio. It looks like we need a good size tank and compressor to get the tacking forces we need.*

Sergio   *That big? Jeez. For that little force we need that big tank? You sure you didn't drop a factor of 2, or of 10?*

Don   *I double-checked. You've got to remember that we want to sustain that force over the width of the bed and for at least a second. A second is a helluva long time.*

Sergio   *And that's what comes out? Did you think of pulsing it? Would that help some?*

Don   *That doesn't make the imaging people very happy. They're afraid of low-frequency vibrations of the bed as it stands now. I think I can maybe make things a bit smaller by changing the angle of the nozzles, but that will mean I have to move the tank back into the feed roller drive. I'm not sure I can get away with that.*

Sergio   *Those guys are going to go bullshit if you ask them to move that drive. It took them three months to get it working right as it is now. Forget it.*

Don   *I don't see much else we can do. I've asked Anand in Analytic Support to take a look at my model. He's a clever guy I hear. Maybe I missed something.*

Sergio   *How about Fred? Is he helping you with the drawing? He's good too. I've known him to be able to shoehorn in a strut or drive belt going places I never thought existed.*

Don   *He is good. But this is a big tank. Fred has sketched out a few possible configurations, but they don't look good. He's still going at it though.*

Sergio   *It doesn't sound too promising. Maybe we ought to drop the whole thing. Jeez, that little force takes that much air, that much pressure?*

Don   *It looks so . . . Ah . . .*

Sergio   *Alright, drop the damn thing. Recheck what you did, but if Anand doesn't come up with something different and Fred still doesn't like it, forget it. Scratch that option. I kind of liked the idea—nothing touching the bed, nothing to mark, wear on it. Bring it up at next Thursday's meeting for the rest of the group. Fill them in on how the air knife is a no go. OK?*

Sergio didn't like Don's story. He had had another outcome in mind when he had added the air knife to his list of options. He had originally envisioned a tank of modest size, a row of air jets that, on discharging, would tack the paper down onto the bed as it moved through the printing process. He had thought in terms of the same fundamental physical concepts and principles that Don had employed, but his thinking was too crude; he was not in the right ball park, so to speak. Don had fabricated a more refined model and a more substantial story that paid full attention to the science of compressible air flow and the laws of momentum and mass conservation. He introduced actual values for pressures, velocities, and flow duration.

Then, with Fred at the drawing board, he explored how an air tank of the required volume, together with the compressor, the valves, and tubing might be located within the existing product. These tentative constructions were not pleasing to the eye. Don's air tank was of more than modest size. His story did not confirm Sergio's original vision.

Sergio, reluctantly, had to give up his attachment to the concept. Once Don had exposed the ingredients and implications of the idea through his object-world construction, and after possible variants on Don's scheme were shown to be fruitless, the air knife became an idea of the past. Don briefly noted his two or three intense days' object-world efforts in his workbook. When he presented the results of his analysis at the Thursday morning group meeting, there was quick and general agreement to drop the option.

Both Don and Beth are working within object worlds. They differ in that Beth's job is to develop a computer model of a working system, while Don's is trying to solve a problem in a system that has not yet been embodied in hardware. Beth must contend with all of the parts of the desalination facility, while Don's concern is with a subsystem for the compression of air and the production of flow through a nozzle—a small piece of the whole system. Beth's hardware is diverse and big: electrical as well as mechanical components that fill up a modest one-story building. Don's vision includes only mechanical elements: a bit of plumbing internal to a desk-sized artifact. Beth's analysis is based on the underlying form provided by the principle of energy conservation, Don's on the fundamental principle that predicts the force due to a change in momentum. And while both bring closure to their accounts, Beth's report validates the design concept at Solaray while Don's conclusion is that the air knife won't work as a fix at Photoquik.

Their efforts are alike in that both have constructed representations of purposeful things that neither of them has seen! That is, they have never been around, turned on and off, smelled, touched, or in any way sensed the stuff of their object worlds as a complete, working artifact. Indeed, that is not essential to their work. They are both able to construct representations—in Beth's case a computer model, in Don's case a first-order paper-and-pencil model—that capture the behavior of these purposeful things and reveal that behavior in hard causal and quantitative terms. They "see" the working artifact through these representations. They connect their scenarios to the furniture of the "real" world through their past experience with actual hardware and through discourse with others, who have still other stories to tell and contacts with yet others, and so on.

To say that they have not seen the artifact demands an elaboration. Beth does see the object in the sense that she has some conceptual image of energy flows and losses; Don sees other imagery of air pressure and a deflected stream of air. These are not things shown on a formal drawing. The air tank is on the drawing; that's what a draftsman sees. This is often taken as the design, but it is not the design. It is a drawing of an air tank. Although a drawing of a cross section of a tank or a nozzle derives from Don's visions and object-world representation, the design is broader in scope and imagery. The design is underlying form as much as a cross section of a tank or a nozzle. We shall see that design encompasses still more.

Both Beth and Don fabricate stories to describe what is going on within their respective object worlds. Beth's understanding and imaginings of energy flow, together with her interpretation of the data from the field and her reading of the performance characteristics of all the diverse components of her system, provide a basis for her computer model. That model by itself is only part of the story; it is just a series of logical statements, in a tightly constrained language, about the numerical values of a variety of factors taken at five-minute intervals. This program and a sketch of a block diagram of the system—all that exists on paper to document her work in progress—require elaboration if they are to carry meaning.

Beth articulates her computer model for others through verbal description. This description includes topics that aren't explicit in the block diagram or in the program—assumptions about continuity of performance of the instrumentation, a feel for the weather in Saudi Arabia, the filtering effect of the sky, the idiosyncracies of batteries, the behavior of the on-site operator, the effect of temperature and dust on the performance of a single photovoltaic module—but are necessary to understand the story. This contextual filling in or backgrounding is a personal activity. Her colleague Tom might have constructed matters with a different nuance, even a different emphasis.

Similarly, Don has a sketch, some relationships scratched out on a page or two in his journal, and references to a text and to some catalogue of pneumatic hardware, but these alone do not constitute an adequate account of why the air knife won't serve. His verbal accounts to Sergio and his colleagues make reference to these markings but are much more elaborate and rich in descriptive detail. They include conjecture about possible alternative solutions using the same principle—variations on a theme—and why these won't work. He speaks of the confidence he has in Fred and Anand; he has cast his net as widely as possible. He articulates and justifies the assumptions he made in his analysis to show Sergio why his initial estimates were off the mark.

It is common for work within object worlds to have this personal character. Exploring variations on a concept, developing and "running" a computer model, constructing a test plan, evaluating a supplier's proposal, or debugging a section of control code is usually a solitary and intense experience. At Solaray, Beth's work was continually interrupted, and she did not welcome these intrusions. But it is not always the case that work in an object world is a solitary activity. Sometimes the object draws the attention of two or three people in consort, as the following vignette drawn from life at Amxray shows.

Digital Designs

Late in the design of the data collection and image processing subsystems of the x-ray inspection demonstration unit, three electrical engineers gathered in a corner office of the company's Cambridge facility. It was shortly before the unit was to be shipped to a demonstration site on the West Coast, and the three were putting together all the pieces to verify that the system would perform as designed. Not unexpectedly, in this integration of subsystems—the data acquisition unit (DAU), the memory buffer, and the data correction and image processing software—some problems surfaced.

The three engineers struggled to set things right. One of them had responsibility for the DAU, another for the data correction software, and a third had oversight of all matters electrical—hardware and software, analogue and digital. They were anxious and disturbed because a test pattern put into the DAU failed to produce the correct image on the monitor screen after signal processing. A check of the values in the memory buffer showed that something was seriously wrong. Here was more than a chance or environmental disturbance peculiar to this particular test; something was awry in the hardware or the software or both.

In the forty-minute discourse that followed, different possible explanations for the fault were entertained: the possibility of a ground difference, a lost bit, a flag set but not reset, a floating state, or an error in the hard-wired logic elements on one of the printed circuit boards. For each possibility, the three constructed a scenario of what went wrong and how this produced the faulty values in memory and displayed on the monitor screen. Each scenario was played out within the participants' object-world imagery and reasoning—image and reason appropriate to the object world of analogue and digital design. Each scenario was fleshed out, given substance for the particulars of this design, and developed as an object-world drama.

Powerful diagnostic instruments were brought to bear. The group leader managed a logic analyzer with the facility of a seasoned technician. Another, responsible for the DAU, set the scope timing and gain with ease.

A spurious blip on the oscilloscope screen drew their attention, and they tried to get it to repeat. They succeeded, and this occasioned a shift in the drama and further conjecture about the path from effect back to cause. An independent measurement was made at another point in the circuitry, and their story held. After they repeated the whole process and observed the same outcomes, their conjecture became explanation, and a fix followed almost effortlessly.

There is an intensity to the brief bits of barely heard dialogue. Everyone else in the office bay senses that the three are not be disturbed. There is no room for casual conversation here, no soft talk about missed deadlines, no banter nor laughter. Mind and hand, thought and object, are wrapped up together. The mind poses an explanation; the object is poked and responds. The object's response is filtered through other objects in the form of instruments that must themselves be put in the right sympathetic condition. The scope measures a small voltage signal over microseconds and, if properly tuned, transforms that information to human scale. The "reading" becomes part of the thought of participants, part of their reconstruction of the object, and leads to revised scenarios and new interventions. We see in this episode how the object infiltrates thought and how thought, reciprocally, configures the object.

In making object-world explanations, different stories are possible, plausible, and appropriate given a single object, and this is the case at any stage of design, even when the process is near completion. I have given examples of explanations made at different stages in the design process: the photovoltaic desalination plant already existed as a prototype; the design of the x-ray inspection system was also well on the road to completion. Nevertheless, different stories are possible.

Words like *story, scenario,* and *fabricate* suggest a flexibility in the object that might strike some as unrealistic or obfuscatory. I don't mean to obfuscate. I want to make myself clear on this point. The object of design, at all stages in the design, is a constructed and contested object in the sense that more than one explanation of its behavior, more than one account, or harder still, more than one analysis of its behavior is possible and meaningful. I mean this in two ways.

First, different participants with different perspectives and responsibilities in the design process, who work within different object worlds, will construct different stories according to their responsibilities and interests. Interest here means technical, professional interest. This is not too difficult to accept. There need be no inconsistencies in these different accounts—they are like ships passing in the night. This is usually not the case, but design is organized to minimize conflict in this respect. Different participants can claim different stories about different aspects of the same object, perhaps at different levels of detail.

For example, the battery is seen by Beth as a black box with certain input/output relationships (a behaviorist's story), while the electrochemist who goes inside the box tells quite different stories about voltages, valences, acids and bases (a cognitive psychologist's story). Or consider this page in front of you. It is an object. A naive empiricist would sense its weight and estimate its size; another reader might note its color or texture; a chemist on the design team would describe its resistance to discoloration, its acidity, and its photosensitivity. A mechanical engineer would be concerned with its tear or its tensile strength, its stiffness, and its thermal conductivity. An electrical engineer would speak of its ability to conduct or to hold a static change. All of these attributes of the object, the same artifact, are understood within different frames of reference, and they all might contend in a design process, for example, in the design of the machinery for the manufacture of the paper, or in the design of the paper itself.

Second, I want to offer a more radical meaning. Even within the same object world and at the same level (e.g., the three electrical engineers discoursing on the DAU and its problem), alternative and different explanations are possible. That is, the explanation for why they saw the erroneous data they did—the explanation that provided them with a rational prescription for a fix that did in fact work—may not necessarily be the only explanation, indeed may subsequently prove untrue or not productive of a solution.

When would it be wrong and an alternative story be required? It would become wrong if at some future time a discrepancy between prediction of performance based on the prevailing account and the observed behavior is noted and made to repeat. All it takes is some unforeseen alteration of context to introduce another factor, a dimension unthought of and latent until now, that would necessitate a new story. It may not happen. But the need for a new and different story to replace or supersede the old is always potentially there.

So, too, in Beth's model. While her elaboration based upon the principle of conservation of energy appears to hold, there remains the possibility that she has missed an internal circuit, or attributed one loss to the wrong cause, thereby making two wrongs a right. So, too, in Don's work on the air knife, if Anand could come up with alternative means to produce a sustained air jet within the confines of the prescribed boundaries and constraints of the existing machine, Don's story would no longer be convincing.

In the working up of an object-world story, there is much testing intent upon falsifying its claims. Once a flaw in the scenario is uncovered, it must be recast and the design redone. The redoing deemed necessary might be minimal and ad hoc or extensive, even to the point of starting over from scratch. Of course, the former is preferred and pursued before any thought of the latter is seriously entertained by participants.

This uncertainty in the validation of a design scenario is what makes designing the challenge that it is. *Uncertain* is perhaps not quite the right word to describe the participants' state of mind, for the authors of these stories display full confidence in their constructions. But it is impossible to carry out all the tests, to develop and pursue all the possible scenarios about the behavior of objects to fully verify a design.[4]

E&M Excitations

The crafting of object-world scenarios and the appropriation of the behavior of design possibilities are not always successful, as this next account of Don's encounter with one of the favored options at Photoquik illustrates.

Within three months of his "options" meeting on the dropout problem, Sergio's team had grown to include Phil, a technician, and a new engineer, Hector, as well as Hans and Don. The electromagnetic option had been proposed by the people in Research—the "cave." The cave was not really a cave or even a basement; it was a building conspicuously far removed from Photoquik's main facilities. The idea was that if you freed up your research staff's attachments to design, development, and production, then researchers—the inhabitants of the cave—would be more creative.

The new electromagnetic option relied on the high-frequency excitation of the bed that the print paper lay on and passed over in the printing process. Don was assigned the task of developing the concept. By experimenting with the hardware and analyzing its behavior, he would develop an appropriately detailed story that would decide the fate of the "E&M option." This he was not able to do. He could not bring closure to his attempted demonstrations of the viability of the concept, and his story remained unwritten, or rather unstated. This is a different result from those of the three preceding vignettes. In each of these there was a conclusion that affirmed (Beth's story) or denied (Don's story) or defined change (the DAU story). These are all satisfactory outcomes and count as successful design work.

At first Don thought that the concept looked very attractive. Preliminary test results from research suggested that there would be few barriers to scaling up the hardware. Some of Don's own initial tests looked very good. He was able to find room within the product to fit the device that provided the mechanical, high-frequency excitation to the bed within, or at least near enough to, the reactant zone. (No big air tank here.) Initial tests looked promising, but in subsequent trials, the positive effect proved sporadic.

Don struggled to bring order to the phenomenon. His goal was to get repeatable test results—not necessarily "good" results (consistent improvements in the printed image whenever the excitation was turned on) but simply consistent results. He sought control over the workings of machinery at his disposal, the furniture of his object world.

The next week people from Product Development changed the machine. For some reason the Atlas Product Development group needed Don's machine for their own purposes. It took another week, once the newer model arrived, to install the excitation device, hook up all the instrumentation, and begin to take data. Not unexpectedly, Don's results using this new setup showed little in common with the last results obtained using the previous machine.

Now this is no simple science lab experiment in which Don can establish control over the phenomenon from the ground up. The context does not allow that possibility. He is given a machine of many dimensions and many variables and must work with that. The paper that is being processed is also variable. Although he tries as best he can to define how change in any one variable changes the appearance of a print, the isolation of causes and effects, with one parameter driving another, is impossible given the resources at his disposal. Add to this the unknown characteristics of the excitation system and the possibility of faulty instrumentation, and the complexity of Don's task becomes clear. Perhaps he was overlooking something that was right there within view awaiting the kind of visual and mental gymnastics that would snap cause and effect into reality.

Clearly some things were more important than others. The parameters central to the stories researchers over in the cave told about the device were primary. There were the resonant frequencies of the excitation source to worry about, and Don did. He instrumented the device, and by measuring variations of current and voltage with changes in frequency, he was able to pin down the resonances. The corresponding mode shapes were a little more difficult to construct.

It quickly became apparent that the positioning of the excitation device relative to the bed was critical. Here things were spatially tight and more difficult to set and vary. It took a bit of manipulation to get to a point where he could claim that the device was located in the "same" position and orientation as in the last test run.

Same brings in the notion of tolerance. The response to the question of what constitutes the same position requires knowing what variation in positioning is permissible. Don, in the process of developing and repeating the test, learns which dimensions are more critical than others and develops a sensitivity of touch in properly locating the device. With time, he is able to do this with hardly a thought—to locate the excitation hardware in the correct position within a few thousandths of an inch. While it would be a mistake to claim that Don could measure "by feel" to a thousandth of an inch, it would *not* be incorrect to claim that, *within this particular context*, he was indeed getting the position right to that exactness.

Two weeks after being confronted with the new machine, Don and Phil, the latter a technician now working part-time with the project, were able to produce repeatable results as they varied the strength of the excitation and the position. The dropout in the printed images appeared to be of the same nature under the same set of test conditions. They were now using a uniform pattern of stars over the entire print as a test image. The outline of a story was beginning to emerge.

Don set about drafting a test plan. He would systematically change three major parameters—the location of the excitation device relative to the bed, the strength of the excitation, and the frequency of excitation—and note the change in the extent of dropout in the test print. He would hold constant all other parameters—the flow of reactants, the print paper feed rate—that might have an effect on print quality. Varying these would engage other interests beyond his control.

Then they lost it. Phil came in one Monday morning, fired up the machine, turned on the instrumentation, and attempted a few standard runs of the reference case. The results were not consistent with prior runs. Don sensed this might happen. He had reluctantly agreed to let Hector sandwich in some preliminary tests on the paper lead-in path change, one of the options Sergio had decided to pursue, late the previous week. This meant some adjustments to the parts of the machine, but these changes should not have affected the workings of the electromagnetic excitation device. Still, Don had enough experience with hardware to know that a machine can perform in mysterious ways. Monday's machine was different from Friday's. He and Phil had to start anew, rebuilding their controlled experiment.

On Wednesday, they were still not able to get repeatable results. At Thursday's group meeting, Don had to report they were not ready to test and evaluate the E&M option.

Don is having difficulties working within his object world, a world of print processing hardware and electromechanical components, a world also of theories of resonant mode shapes and frequencies, of instruments measuring voltages and currents, of power supplies and frequency generators, of paper and all the chemistry of the complete print process. His work within this world is conditioned by social and organizational events. He must connect with and rely upon the advice of the research staff; he must learn from the suppliers of the excitation hardware about its undocumented intricacies; he must negotiate with members of the Atlas product development team—pressing for the option to make slight changes in the current design. And he must deal with the needs of his peers working under Sergio's direction.

Don does not know his object world well enough to make it produce repeatable data. Remember, too, that he is new to this branch of the firm. The elements of his object world won't sit still in a respectable order. He claims to understand the underlying form of the E&M hardware, but evidently his grasp of fundamentals rests at an abstract level too far removed from the ways of this particular device in this particular context. The totality of the furniture of his object world—excitation hardware and suppliers' specifications, voltages and frequencies, mechanical drives, theories about modes and frequency of vibration, sites of blemishes on test patterns, positioning tolerances, print paper weight, product constraints, and Hector's doings—remain disconnected, like shards of a puzzle lying about, scattered in his mind. He freely associates at the slightest prompting, making this combination or that, attempting to explain the aberrant behavior of his object. Don constructs scenarios, but they do not hold; they don't prove capable of explaining the variety of results he and Phil have observed in running the variety of tests to date. Has he missed a piece? Has he gotten his priorities straight? Is he being too crude with respect to some parameter? Don's world is in disarray. He struggles to bring it into order. He has no story to tell.

Varieties of Object Worlds

These vignettes about Beth's model of a solar-powered desalination plant, about Don's evaluation of the air knife, about the trio at Amxray testing a subsystem of the x-ray inspection machinery, and about Don's struggle with the E&M option show different individuals working within their respective object worlds. We see them applying the fundamental concepts and principles of their disciplines, manipulating hardware, analyzing, sketching, instrumenting, testing, modeling, and constructing scenarios describing the way things work, or ought to work, in their struggle to gain mastery over their designs. Each labors to make the object his or her or their own, to appropriate its behavior in the particular situation at hand. Each mulls it over in thought, makes working representations and tests them in the laboratory, formulates computer models, and exercises and probes notions in discourse with colleagues.

In this, each individual brings to his or her object-world deliberations a personal rendering of scientific principles and technical possibilities. We can even speak of different styles, of tacit knowledge or of personal knowledge, and to do so is not soft but necessary if our intent is to understand how participants in design add flesh to the bones of an idea and how ideas are embodied in particular hardware or systems.

Though tacit or unarticulated, an individual's personal knowledge is critical to the quality of the design. The "feel" that designers bring to their work is part and parcel of the know-how that enables them to bridge the gap between the formal, abstract knowledge of underlying form and the practical concreteness of the immediate object. At Beth's firm, a colleague responsible for overseeing crystal growth of the photovoltaic raw material displays a sense of proper rate of growth and of the nuances of one growing machine relative to another. An electrical engineer in Tom's group shows an understanding of and concern for leakage to ground in a network of cells that is not documented in any text. Tom himself talks about "load paths through a structure," and the necessary ingredients of an efficient structure, as if he were describing his favorite recipe for a white sauce. Even the firm's comptroller displays an intuitive grasp of the ever-shifting status of inventory.

These different facets of different participants' object worlds are usually not made explicit in formal technical presentations, nor is their form prescribed by any set of rules. But this kind of tacit understanding is critical when a real object is the focus of attention. The singular shows its own quirks and special features and demands tailoring of generic knowledge, no matter how sophisticated or detailed, to fit the special circumstances of the immediate design task.

Within a single discipline, say within the world of a mechanical engineer, different individuals can show a variety of styles and tacit knowledge. These differences may derive from differences in education or national origin as well as from the particulars of the enterprise within which they labor. Take the concept and practice of tolerancing as an example. One mechanical engineer's knowledge of an appropriate tolerance may not be the same as that of a colleague, even though they work in the same bay. What "±0.002" means to an individual depends upon the person's past experience with specifying tolerances and with particular forms of machinery machining particular kinds of artifacts. What at first sight appears as a single, precise number on a detailed design drawing signifies, within the designer's object world, a personal understanding of common machine-cutting practice, the kinds of processes appropriate to achieve different shapes and surface finish, and some sense of the relative costs of these different techniques as well as why the design may require that particular tolerance in the first place. How much variation in a dimension to allow, what variation can be tolerated, is a matter, in part, of instrumental "fact"—statistically we can expect a three-sigma deviation of 0.002 inches—but it is also a matter of judgment, of knowing the quality of the work of this or that particular supplier.

This is one example where the tacit, the experiential, weighs in. Another is the system of units engineers use in working within object worlds. There was a time when the United States was going to convert from the English system of units to the metric system within a specified number of years. That has not occurred, although the time frame has elapsed. It has proven to be a more difficult transformation than anticipated, more than simply a matter of a straightforward numerical conversion from inches to centimeters, from pounds (force) to kilograms (mass), from Fahrenheit to Celsius. People in charge recognized that it would take time. They knew that things like standard screw sizes and the specification of standard dimensions were expressed in inches and frozen into the hardware, so to speak. Still, professional engineering societies went on a campaign to move their members to "think metric." It hasn't worked out; the societies' expectations have not been met.

Why? Because to think metric is more than a numerical conversion. It is a form of personal knowledge, including a feel for an inch, or foot, or pound that is not there when we speak or write "meter" or "kilogram." Of course, we can stop and figure it out, look it up in the tables, or punch it out on a calculator, but the bit of time (seconds) it takes to make this simple numerical conversion is like a gap across a canyon separating two different worlds. We don't know metric the way we know English. It is not part of our culture, our object-world heritage.

A Final Example

Ed at Solaray had contracted with a consultant, a professor of electrical engineering at a neighboring institute of technology, to conduct an analysis of the solar heating and cooling of a remote facility—a relatively small service island in the Saudi sun and sand isolated from the utility grid. The plan was to use a photovoltaic system to provide power for lighting and for air conditioning. The service island would be occupied twelve months of the year, and the comfort of occupants needed to be assured. To size the photovoltaic system, Ed needed to know how much power and energy it would take to run the air conditioner, which was by far the major load. More specifically, he needed to know how the demand for power would vary over the course of the day, from day to day and season to season. To obtain this information, he needed a model of the solar heating of the building during the day and cooling of the facility at night.

He had been referred to Professor P. by a government laboratory that had used P.'s computer simulation of the heat transfer through buildings and had been pleased with the results. P. came out to Solaray, gave Ed a version of the simulation software that he thought would work on the Vax computer, and instructed him in how to model the facility. Ed found the input format easy enough to appropriate, and he had the simulation running in due time with a little help from the in-house computer systems manager; but he didn't like the form of the results. Not that they looked incorrect, but the time history of heat flows and energy flows was varying much too slowly. Again, it wasn't wrong; it was just that there was a superabundance of data, changing in value very slowly from one instant to the next, over the course of a day. It was like watching a movie of a dynamic event in ultra slow motion.

The program also took what seemed to Ed to be an extraordinarily long time to run through a case. Because he had many cases to study, turnaround time was important. He was running the program interactively, sitting in front of a monitor, and so delays in waiting for a response to appear on the screen were amplified to a bothersome degree, even though each delay was measured in seconds.

So Ed, against the advice of some, decided to dig into the program and see if he could speed things up. He sought to appropriate the tool to his own use. He knew that the computation of the energy flows is done in chunks of time; the temperatures at each instant depend upon the temperatures at the previous instant and the energy flows over the chunk of time between the two instants. For the program to work efficiently yet still produce sufficiently accurate results, the time increment must be chosen correctly.

It didn't take him too long to find the problem. Professor P. had set the time step as one second. To remedy the situation, Ed simply changed one line in the program, increasing the time interval to one minute. This did the trick. His results looked a bit different from the previous runs, but the difference was insignificant. His output was reduced by a factor of ten or so, as was the run time.

There is a strict analogy between heat transfer and the behavior of certain electrical circuits when that behavior is described mathematically. The same formal, abstract system of (differential) equations can represent the underlying form of these dramatically different physical phenomena. Physically, they are the stuff of at least two different object worlds: heat transfer is the province of the architect and the mechanical (or chemical) engineer; electrical circuitry is the domain of the electrical engineer. Yet the language of mathematics, so essential to all object worlds, appears to make them one in essence. And it does, in principle. The two disparate phenomena have the same mathematical structure.

Because they have this common underlying form, they are good candidates for illustrating variations in personal knowledge associated with life within different object worlds.

In the world of the electrical engineer—Professor P.'s domain—things like capacitors, resistances, and inductors, when assembled in a circuit and activated, can react dynamically in milliseconds, but circuits with "time constants" on the order of minutes or hours are rare. The opposite is true for the mechanical engineer concerned with the heat flow into and out of a building in the Saudi desert. It takes minutes and hours for the temperature to change significantly within a well-insulated building. Professor P.'s choice of time step belonged in the world of electrical circuitry. He thought of one second as a *very long* time step, appropriate for heat transfer.

Even though the program functioned correctly, in accord with underlying form, the time step was inappropriate. It wasn't in error or wrong in a mathematical sense; there was no violation or cutting corners in modeling here. And, indeed, one could argue that Professor P.'s choice of time step was better than others he could have made considering that too large a time step could lead to inaccurate results. But it was wrong in the sense that it was the wrong color, wrong in the sense of going to Fenway Park in a tuxedo. It was wrong in that it was attuned to the specific context of another object world. Ed, in opening up the box and grappling with the inner workings of the tool, appropriated it for use within his own object world.

From the perspective of scientific rationality, Ed's change of the time step appears insignificant. Indeed, to talk of tacit knowledge appears trivial and inconsequential from the perspective of scientific rationality with its purported avoidance of all that is personal and stylistic. But from the perspective adopted here, Ed's action is a significant instance of object-world activity critical to designing. His appropriation of the object is essential to his work. His story about the energy requirements of the service island and how they will be met by the photovoltaic system carries authority because of his investment in opening up the program and mastering its contents. It enables him to convey to others his understanding of the heat transfer process and its implications for design. If a colleague were to ask him to increase the amount of insulation, add another series string of batteries, he would show the robustness of his understanding in his response. The object is his to control, to manipulate, in his mind, on the computer and eventually in hardware on site.

Object worlds are hard in the sense that their base content is "objective," instrumental, often formal and abstract. At the same time, object worlds are personal worlds. They derive from an individual's schooling in a discipline, are tempered and shaped further by an individual's work experience, that is, his or her professional history. Within these shifting contexts, the individual is the agent of their construction. While they are shared, they remain variegated; they are given new expression and show a different nuance from one design task to another. Learning through design action reflects back upon the content of object worlds, stimulating new interpretations and implications as well as reinforcing the core concepts and principles of these worlds.

Let me simplify and summarize. A mechanical engineer works on a design task within a world of mechanical contrivances, concepts, and principles; an electrical engineer on the same project bears responsibility for voltages, currents, and the like. One object—say a photovoltaic module—is part of different object worlds. Now we go further and claim that different individuals working within the same discipline will construct personal, and different, renditions of the object. The most important distinction is that among different disciplines, but the second kind of distinction must be allowed if we are to explain how design outcomes are successful in some contexts, not so in others, given roughly the same design task.

Yet expertise and competence within an object world are not enough to ensure quality design. All participants must be able to explain and describe their experiences to others from different object worlds who do not have the same depth and expertise or the same familiarity with underlying form and its technical possibilities. Despite differences among individual interpretations and constructions and among object worlds, participants do communicate, negotiate, and compromise; in short, they design. That they are able to do so suggests a common structure, shared by all participants across object worlds, for patterning explanations and fixing what counts as an explanation of consequence and what is relegated to "background" noise. This framing is the subject of the next chapter.

4

Cosmology

The Rhetoric of Objects

Work within object worlds is only part of the engineering design process. An individual may concentrate on exploring a concept, "debugging" a subsystem, or modeling and evaluating a prototype system in the field, and the descriptive explanations he or she produces will be essential to the design task; but these individual productions must be brought into concert with those of others for design to proceed. Contemporary design is, in most instances, a complex affair in which participants with different responsibilities and interests—that is, working within different object worlds—must bring their stories into coherence. This is no simple synthesis achieved according to some straightforward instrumental technique, as much as that might be desired by professors of management or engineering design, or even by the participants themselves. Object worlds are not congruent. Interests conflict, trade-offs must be made among different domains, and negotiation is necessary. Design is a social process as much as it is getting things right within object worlds.

The framework for common discourse across varying object worlds is a composite. More than a common vocabulary and syntax, it is a web of tacit understandings of what is to be considered an honorable claim, a significant conjecture, a valid "proof," or a laughing matter. It is an accepted rhetoric for describing, proposing, critiquing, and disposing that girds all design conversation, fixing what constitutes a true and useful account. What follows is an attempt to define the ingredients of this rhetoric.

Object-world reports have a directness and sameness that reflects their required adherence to some universal theme; but at the same time they can be rich in their detailed elaboration of underlying form in terms of the particular object at hand, and their construction requires energy and creativity. Once closure is attained, however, they become fixed, repeatable, and immediately less interesting. For example, once the group at Amxray has fixed upon a scenario explaining the fault in the performance of the DAU, the drama is over, all uncertainty and ambiguity have been locked out. At least this is how matters are perceived. Similarly, when Beth has accounted for all the energy flows along all possible paths in the desalination plant, she can be assured that she has a consistent model. Object-world thinking is thinking about the rigidly *deterministic*. This is its primary characteristic.[1]

The stories everyone struggles to construct are also about the *abstract*. Reductionism is rampant. Models of the sort prevalent in science, and often derived from science, are essential to work within object worlds. We have seen this in Beth's model of energy flows and in Don's talk about momentum and force in his model of the air knife. Tom, as an electrical engineer, sees the ensemble of cells within a photovoltaic module as a network of ideal current generators connected in series and parallel and forming a pattern whose topology is critical. All elements within this network—diodes and photovoltaic cells and the like—have their own abstract and idealized electrical properties, which relate according to well-defined abstract and general rules. From these circuit relationships, given the topology, Tom can estimate the amount of power the network will produce for a given intensity of solar radiation falling upon the module.

When Ed down the hall confronts the task of specifying how thick the glass that protects the network of cells must be, he thinks of the internal stresses (another abstraction) generated within the glass by, say, wind loading. He brings to mind a mathematical theory about stresses and strains in a continuous thin plate, whose relationships (abstract) allow him to estimate how thick the glass must be to keep the internal stresses below a certain level.

Participants reveal the totally determined character of the stuff within object worlds through *cause-and-effect* chaining of the agents in their stories. The sun's radiation impinging on the photovoltaic module causes a flow of electrons through the thin cell. The current is channeled by the network of connectors to the junction box and then to a "load," which could be a radio, a water pump, or a fluorescent light, and it is further transformed there to useful purpose. This causal chain of events and transformations is, in theory, capable of being explained to any desired level of detail—in this case, to the level of stories about why and how the electrons flow through the cell, or why and how the water pump pumps, or how electrons making transitions within a charged gas give rise to the glow of a fluorescent light.

Object-world stories work better with fewer elements; abstraction and reduction go hand in hand in this business. Sparseness characterizes a good, workable model. But the work of design also demands a fabrication of detail and an expression of underlying form in terms of worldly furniture and hardware. Reduction and abstraction in terms of concepts like energy, momentum, force, voltage, and current is just one aspect of object-world activity. Participants' stories are also about the *concrete*: the open-circuit voltage of a particular photovoltaic module, the diameter of a supplier's air tank, the response spectrum of a particular lot of crystal detectors.

The connections between the abstract and the concrete are set in the story-making process and are normally expressed in *measured terms*—literally. Participants talk about the current output of a photovoltaic module measured in amperes, the energy it delivers over the course of a day in kilowatt-hours, an air tank's diameter in inches. But these quantities are not in themselves so immediately concrete. Interposed between the engineer and the thing in itself is an instrument that makes the measurement. This is usually transparent; and the identification of the quantity seen on a display or dial with the abstract entity is taken as immediate and synonymous. When the object misbehaves, however, a technician may admit the possibility of instrument error and open the black box to consider its role in the production of the absurd.

Designers' stories are about things that are made to be measured, but their reports may not always be rendered in precise quantities. Often ball-park *estimates* and anticipated ranges will suffice.

Object worlds are closed and finite; within them, *conservation* reigns supreme, setting bounds and limits on the measure of things that can then romp over the entire domain. Conservation can be the main theme of an account, as in Beth's use of the conservation of energy, or it can be peripheral, limiting the excursions of a single measure, as in Don's realization that there is only so much space available for the air knife. Conservation is the basis for all measurement systems within object worlds in a pragmatic as well as an abstract and lawful sense, setting scales and resolution as well as range.

Note that the boundaries of a domain must be specified if a conservation principle is to be of any use. Boundaries are critical in design. If the budget for a project is unlimited, designing is fruitless, well nigh impossible. Constraints are as necessary to design as they are confining.[2]

So, too, conservation applies in decision making. It finds expression in the way participants set limits on the number of options they will entertain. It is there in phrases heard in jest: ". . . no such thing as a free lunch" stands as a projection of object-world norms out onto the world of process. It is there in discussions of the projected costs of an engineering change order, a slippage in schedule, or whether to go with a higher-quality fastener: "Life is a zero-sum game." "You never get out more than you put in." One has the sense of a set of physical objects lined up to be counted behind these lawlike statements. In object-world talk, time, people, and money are the same things as energy, mass, and momentum in the sense that all are finite, measurable, and bounded.

Object worlds are structured worlds. Participants in design see *hierarchy* as essential within them. This is reflected both in their conceptions of underlying form and in the organization and doing of their work. Nature appears to have structure and shows a hierarchy, which is reflected in levels of scientific abstraction. The general relations that hold among the forces acting on a continuum in static equilibrium are more "fundamental," hence higher up in the hierarchy, than the equations derived from them that determine the stresses in a cantilever beam. The theory in the solid-state physics that describes how radiation from the sun stimulates the movement of electrons in a thin, doped layer of silicon is more basic than the model of the behavior of a photovoltaic cell found in a text on circuit theory. The hierarchy of underlying form(s) is like a Russian doll that contains a nested set of dolls of decreasing size. At the core sits the kernel of nature's truth, the root form of the hierarchy.

Beth, Don, and the trio at Amxray must know the correct level, both of the abstract and the concrete, at which to work. Beth treats the photovoltaic module in her simulation as a "black box."[3] She need only be concerned with its "performance" as defined in terms of the current-voltage characteristic curve and how this shifts with temperature and the intensity of solar radiation. She knows she need not be concerned with the performance of each individual cell or the physics of how the cell transforms solar radiation into a movement of electrons (a current). Similarly, the trio debugging the DAU take the instruments they apply for granted, though these in themselves might be considered a possible agent of the "bug." Don leaves to Fred, an experienced mechanical designer (in the narrow sense), the task of depicting the shape and location of the air knife apparatus. His concern is with the volume and nozzle dimensions as defined by his fluid-flow analysis—the few abstract and concrete agents that dictate the play of his story. These instances all reveal the respect for hierarchy in the work within object worlds.[4]

A belief in hierarchy and conservation frames the construction of trade-offs in a design—structured, binary decisions between options made on the basis of a comparison of relative strengths and weaknesses. This is seen as a measured act in which an "optimum," or at least a "satisficing" solution,[5] subject to the constraints of the problem, is attainable.

Along with these common facets of object-world stories comes a structure for the account itself—a common pattern or format and way of reporting. The decisive, hard, technical, instrumental edge to these characteristics of object-world accounts—abstract, quantitative, lawful, bounded, hierarchial—suggests that the accounts are just a theory of operation, the sort of account that is available in a technical manual. After all, they are all about the inanimate—for example, Beth's story about a photovoltaic array and the energy it produces, about batteries and their efficiency, about a control system and how it keeps the batteries from overcharging, about pumps and filters and the like.

But "theory of operation" doesn't capture the full nature of what Beth must construct if it is to serve as an evaluation and basis for design. Her description must have more punch and vitality; it must be more than a description, more than a theory of operation. It must suggest a direction for change and improvement. Indeed, in some ways her story will display the features of a human drama with a moral to be learned.

As an alternative, one might propose "puzzle solving" in place of "theory of operation." After all, these vignettes are about individuals figuring out why things don't fit, trying alternative explanations, musing and searching for the piece that will lock in an ensemble of features. Then, too, puzzle solving is usually a solitary and intense activity; the joy of finding a missing piece is akin to the engineer's delight in finding the bit of evidence that supports his or her latest model or verifies the functioning of a prototype.

But the puzzle-solving metaphor won't do either. A puzzle has a unique solution in the form of a specific arrangement of pieces, even if there many ways to arrive at a solution. Puzzle solving is getting all the given pieces to fit together in a predetermined pattern. This is not the case in designing, or else it's a very peculiar kind of puzzle that Beth and Don and the others are working on—one in which the pieces need to be constructed as they go along, the boundaries may not be fixed or straight, and the image of the whole may not predetermined in all of its features. No, puzzle solving, although attractive, won't do.

"Story making" is a better metaphor—story making about voltages and currents, tank dimensions and air pressure, vibrating crystals, digital data bits, and the like. This story making is constrained by these mundane concepts and features in that they must behave deterministically in accordance with an accepted set of principles and rules, and there is general consensus not only on what they are (voltage, stress, pressure, flow rate, etc.) but also on how to measure their character in hard, quantitative terms. Nevertheless, there remains much to be constructed of their relationships and interactions in particular circumstances within object worlds. The elements are there, but their synthesis requires rhetorical skill and creative effort.

One might suppose that the language of Beth, Don, and the trio at Amxray working within their respective object worlds would be as precise, as instrumental, and as constrained as that of the formal texts in mathematics, science, or engineering.[6] This is not so. While the laws of science and their implications for engineering are intended to relate precisely defined, abstract concepts in a dispassionate way, in this "correct" form they are so far removed from the particulars of their instantiation within object worlds—worlds of pumps, crystals, batteries, relays, connectors, and voltage levels—that a good bit of creative effort is required to construct a coherent and useful story about how they work together.

There is a considerable flexibility in the ways Don or Beth or the trio at Amxray might appropriate an object and make it their own. In their personal rendering of the specific object at hand, the richness of metaphor and language allows each individual to gain control and to master the object on his or her own terms. Of course, this is not just an "anything goes" intellectual exercise but demands the use of intermediate forms, such as computer models, testing down in the lab, data from the field, mock-ups, and the like—as much as can be accommodated given the resources available. These efforts require creative renderings of scientific concepts and laws, and they also depend upon contextual as well as formal knowledge. Thus, while there is a necessity to the relationships underlying the "behavior" of objects within object worlds, so much remains to be fabricated that, to the author of and audiences for these stories, their ingredients and character show an undetermined and plastic, indeed vital, form.

The language of designing is accommodating. Listening to these stories, one would think that the teller was describing a dramatic event in a television soap opera rather than the daily excursion of a photovoltaic array current, the falling out of a digital bit in memory, or the gross situation of a plenum filled with compressed air:

Sergio   *Jeez. For that little force we need that big tank?*

Why is Sergio so excited? After all is said and done, the principles of science correctly applied determine the behavior of the air flowing and its pressure within the tank. All is fixed. Indeed, to speak of these episodes in a dramatic fashion is philosophically suspect, although the metaphor may ring true enough. Sergio talks this way because he knows that the objects of these stories are not so mundane; their character can be voiced in different ways, the rules they must abide by can lead to more than one tale and embodiment of principle. Another story might still be written about air pressure and tanks and forces that might lead to a different denouement.

We can draw a parallel with the poetics of Aristotle: Even though the agents in these object-world stories are well known, like stock in trade, and even though the general course of their relationships is "given from above" (dictated by science), participants in design still must construct and develop both plot and character appropriate to the setting of their particular design task.

All of this is in "imitation of nature, and man's productions out of nature." That's what a model is. But it is more than "a mime or a simple tracing." It is "artful imitation"—a representation that the audience can learn from, so that they will understand the playing out of underlying principle in the effectiveness of a photovoltaic-powered desalination plant, an x-ray inspection machine, or a photoprint processing unit. At some abstract level, the plot is well known in each instance, yet many different dramas can be written about the same artifact—one artifact, many object-world dramas—"but it is still left to the poet himself; it is for him to devise the right way of treating them."

"The causes of the actions and interactions that make up the plot should be included in the episodes of the story." The task of the designer is to make evident physical causes and their consequences. It is a bridging activity: To take a transcendent law and tie it down for a while in an episode, as in the narration of the behavior of the desalination plant. "The action is to appear necessary or as the probable outcome of the character of the agents" (currents, resistances, flow rates), which "must act within the confines and dictates of the plot so well known."

Note the distinction between this model making and a historical accounting. Beth's model, for example, although constructed using data about the behavior of a photovoltaic system working in the past, is not historical verse; her model describes more than "the thing that has been." It is "something more philosophic and of graver import than history, since its statements are of the nature rather of universals, whereas those of history are singulars." Beth's model of a photovoltaic system will account for all such systems, for all time. History is contingent; the object-world model holds for all time.

Here, then, is a way to view what participants in design are about when they are working up their design concepts, evaluating a particular configuration, or proposing a system to perform as the request for proposals demands: In their modeling, evaluating, and designing, they are constructing stories, but it is a constrained and formal kind of story making, in accord with the strictures found in Aristotelian poetics.[7]

More or Less than Meets the Eye

Drawing is an important ingredient of the language of design. Sketches, graphs, mechanical assembly drawings, circuit topographies, block diagrams, and charts of all sorts are key features of participants' accounts and productions. Drawings, in this general sense, show the characteristics displayed in narratives and, indeed, are themselves essential to narrative. They show hierarchy, are abstract, bounded, measured, and so on. These are not just characteristics of the formal drawings stored and saved for posterity (about which we will talk first), but they also structure the hastily rendered sketch made to assist in the storytelling of the moment. These latter, like the narrated stories themselves, rarely survive for future generations to inspect.

At first glance, the formal productions of designers, such as the drawings that inform manufacturing how to fabricate a part or assemble a system, appear complete and unambiguous. They are strictly Cartesian in the sense that within them matter fills all space. No two objects can be in the same place at the same time, and all relative positions must be clearly coded. The value of a final detailed design drawing is in how well it makes clear how objects fill space and relate one to another. Visual ambiguity is error. An Escher drawing would not do as a plan.

In making a formal design drawing intended to show how material artifacts are to be assembled, designers must ensure that the pieces can be put together and then the whole put into use without causing awkward moments or contortions on the part of the artifact or its owner. Unplanned interference among the pieces is failure. That is why designers must specify tolerances on dimensions: No machinist's cut is perfect, no extrusion runs as true or as straight as Euclid's line connecting two points.

Formal drawings of parts and their assembly appear complete and clear, but they are in fact an abstraction. Consider the detailed drawing of a part. Its spatial representation is abstract in that only essential dimensions are shown. Redundancy in dimensioning is to be avoided, and there is a well-established convention for representing the fully three-dimensional artifact within the two-dimensional object world of the paper. Both author and reader must know the underlying principles of orthogonal projection if they are to read out of the drawing its three-dimensional features.[8]

And while the formal drawing will show surface finishes, tolerances, and special effects as well as the spatial arrangement of parts and their dimensions, some information necessary to the part's production will not be shown. For example, the choice of method for machining may be left to people on the shop floor, and some dimensions will not specify a tolerance, indicating that a default value is to be assumed. Details are critical to the construction of an acceptable formal drawing, but just which details are shown is as much a matter of local conventions and understandings within the firm as it is of globally applicable principles found in textbooks on engineering drawing.

The abstract character of formal drawings becomes clear when we move to nonmechanical object worlds. A formal drawing defining a network of cells within a photovoltaic module is meant to be as clear and unambiguous as the companion drawing showing the spatial layout of the material cells within the module, but scale and dimensions matter little in this case. Like the mechanical drawing, though, this one is also Cartesian in the sense that the network of current sources, diodes, capacitors, and resistors must be topologically complete and accord with established conventions for showing the underlying form of circuit theory and the behavior of electrical devices. In this sense there are no spaces left undefined or empty, and the whole is bounded. Knowing underlying form and how it is to be represented is prerequisite to reading the drawing, indeed prerequisite to making the drawing in the first place. A drawing's features and their meaning can only be understood within the context of its abstract nature; the symbols within an electrical network represent the ideal behavior of the devices signified, and the circuit itself is a spatial representation of abstract law—Kirchoff's laws—as underlying form.

As in the case of a mechanical drawing, the formal representation of the workings of an electrical network might be considered incomplete if it doesn't indicate the style of a connector to be used at an interface, the spatial routing of all leads, and other material features of the finished product. From this broader perspective there is much that is missing. (The particular type of connector may be presumed; the cable routing problem may be taken as trivial.) From the perspective of those working within the narrow confines of the electrical network object world—for example, the engineer responsible for producing a diagram that describes the function of the circuit—the additional information is irrelevant, like the ivy on the remnant of the wall supporting Galileo's beam. Yet others, or even the same engineer on another occasion, could very well be concerned with the material embodiment of the electrical network and the devices it contains. Clarity, completeness, and the Cartesian nature of formal drawing only have meaning relative to a particular object-world context.

When we turn to the informal images that are generated more frequently by participants in design, the significance of context again becomes evident. Figures 2–5 show four sketches made in the day-to-day work of design.[9] They are diverse; and while they do not represent all types (notably missing are images of three-dimensional artifacts), they do give a sense of the kinds of graphic imagery that are most frequently encountered in the design process.

Figure 2 is a hand drawing of an electric circuit that shows not just a circuit topology, but the relative spatial location of all leads within the network. It is really two drawings in one. The solid lines represent the circuit on one side of a printed circuit board; the dashed lines, the circuit on the other side of the board. It is a rough drawing—an interim construction that will be used by its author and other participants in design as a basis for a set of more exact, computer-generated drawings that, in turn, will be used to create four masks for etching the two sides of a printed circuit board. It is a personal sketch, liable to be altered and improved upon in this process. Although it shows a certain aesthetic quality, it is essentially lifeless outside its intended context.

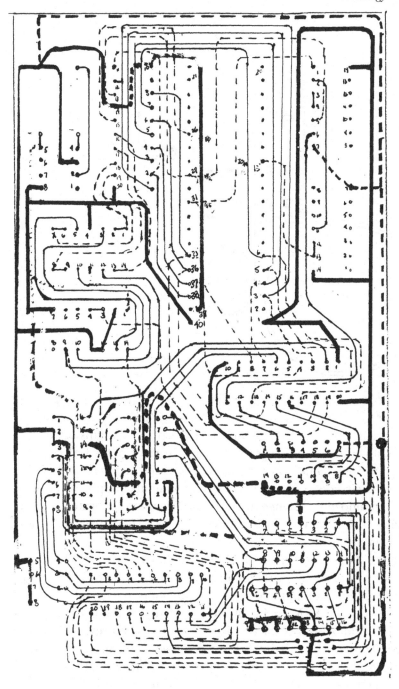

Figure 2

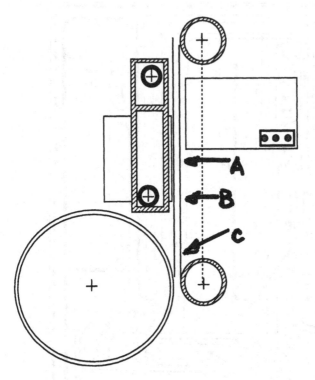

Figure 3

There is a set of meanings encoded in the relative widths of the lines that both author and reader must know. The thickest lines carry power to the various "chips" on the board. The chips themselves are shown only by their "footprints," which consist of a number of pin attachments (some multiple of 2). Since there are no dimensions indicated, one can assume that a standard etch width is implied. Hole size may be also taken as standard.

Figure 3, the only one of the four that shows mechanical hardware, is an overlay of pencil markings on a portion of a more formal drawing. It was made by Don in the course of his attempts to determine the best location for the E&M excitation device relative to the bed and paper feed of Photoquik's Atlas product. It is thus a representation of a physical system. But it is not a spatially complete view, in that the formal conventions and rules for depicting a three-dimensional object on a two-dimensional sheet of paper have not been rigorously followed.

It appears as a section view of a portion of the photoprint machine. Its main purpose is to show the three locations, marked A, B, and C, where the E&M excitation device might go. More precisely, we should

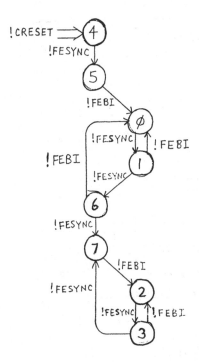

Figure 4

read it as the superimposing of three views, each showing a different location of the excitation device.

Is this an accurate representation of this section of the machine? Some hardware units are shown, and these we take as faithful representations. But note that you could not fix the three-dimensional form of these units from the information provided. The image differs from a formal drawing in that all the details of the true section are not included. It is an abstraction showing only information critical to locating the excitation device. Some extraneous features function as a chorus in the background, serving as stage setting and orienting the reader to the machine.

The figure presumes a reader thoroughly familiar with Don's object-world efforts. The spatial presence of the drive roller is evident but only in two dimensions; the tightness of fit of the various components is indicated, but only someone who has opened the front panel of the machine, released the bed, pulled it out for servicing, and then readjusted the feed roller would see this. Phil, the technician working with Don, has no difficulty reading the amended drawings.

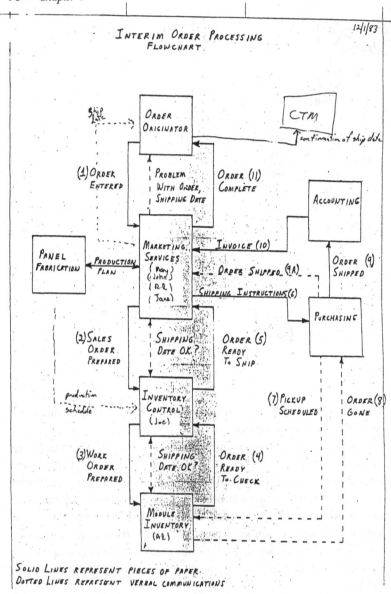

Figure 5

Other details important to the full program of testing and developing the E&M option are not shown but may be assumed to be there somewhere. Examples include the cabling of the leads to a power supply and the details of the fasteners holding the device in place. Such details are of secondary importance and will be understood as such by the parties to any exchange within which this sketch is deployed.

Figures 4 and 5 are similar to each other (and different from the first two) in that they are meant to capture events that change with time. They show some enclosed regions, circles in figure 4, boxes in figure 5, connected by lines. The lines have arrowheads on them. In both, the enclosed regions are labeled and represent the steps in a process through time. Figure 5, which depicts an interim order processing flowchart, will be the subject of an extensive analysis below. Figure 4 also depicts a flow of information, but in this case information rooted in highs and lows of voltage signals and their "values" as bits configured as "octal word" bytes.

Figure 4 shows fewer elements and appears simpler, but its intent is the same as that of figure 5—namely to represent the states of a dynamic system whose ingredients are changing with time—as in Beth's computer model of the desalination plant's performance, except that the time increment is different from Beth's or that of figure 5. A shift from one logic state to another, deep within a computer, happens in a matter of microseconds; a significant change in the power output of a photovoltaic module takes minutes; and the time scale of the interim order flow chart is marked off in hours and days. Both figures 4 and 5 make static what is dynamic and presume that the reader will understand, without bringing to mind, the appropriate measure of time. That measure does not appear explicitly in the figures.

We can think of these sketches drawn in the course of object-world activity as "speech acts," as part of the process of making and telling stories.[10] Some are worked up and become permanent ingredients of the design; others are more transient evocations. There is little in them that tells you about their significance and role in the negotiation of design options, what followed the discourse of which they were an integral part, or how they stimulated their authors or audience to adjust their own thoughts and practices. Note, too, that one and the same drawing may be used on more than one occasion, layered with new meanings, shadings, comments, and erasures. In this case it might be best understood as an entirely new drawing.[11]

The correct reading of all of these drawings requires a knowledge of the local dialect of the object worlds to which they belong and also of the context of their moment in use.[12] Whether they represent spatial configuration, a static topology, or the dynamics of a flow process, they are sparse and abstract, symbolizing the essential features of whatever it is they are about. The reader who is attuned to the hierarchy of concepts and forms within their respective object worlds will not notice their abstractions. Foreground and background, up and down, input and output, why one line is dotted and another is solid—all this should be clear in the moment of seeing.[13]

Although many details are left out, the important object-world content of these figures is there in relief. All show deterministic configurations in space or time. If alternative routes are possible in a flow diagram, these appear explicitly as options. All possible trajectories or states are included in the representation. All causes and effects are displayed. There is closure, and the system is bounded. All is clear, unambiguous, and certain, at least if one is capable of right thinking, reading, and speaking within the relevant object worlds.

Roots

"Right thinking" is not inherited. It is learned, first in school and then on the job. To get a fix on roots, we need to step outside our studies of process to probe object-world thought and practice as woven through the preparatory studies of university students. Rather than attempt a comprehensive survey of engineering curricula, seeking their implicit and explicit influence on the way participants in design think, believe, and value, I will analyze a prototypical experience of the undergraduate years, one that may occupy 80 percent of a student's time—namely, the solving of problems.

Problems, most admitting of only one answer, are the stuff of quizzes, homework, and final exams and hence determine the student's grade. I have taken one (almost but not quite at random) from the third edition of James Meriam's *Engineering Mechanics: Statics and Dynamics,* a text first published in 1966 and well known to faculty of engineering.[14] I will first focus narrowly, trying to understand what might constitute a "correct" reading of the problem, that is, a reading that enables the student to "do" the problem and get the right answer. I will then stand back and, taking stock of that process, ask what the student has learned about ways to perceive, to value, and to neglect while on his or her way to a solution. I will ask, in short, what is really going on when a student confronts and struggles with this exercise. What frame of mind is prerequisite to a successful encounter? What resources must the student bring to the task, and what must he or she leave behind?

Here is the problem:

The hydraulic cylinder gives pin A a constant velocity $v = 2$ m/s along its axis for an interval of motion and in turn causes the slotted arm to rotate about O. Determine the values of $\dot{r}$, $\ddot{r}$, and $\dot{\theta}$ for the instant when $\theta = 30°$. (*Hint:* Recognize that all acceleration components are zero when the velocity is constant.)

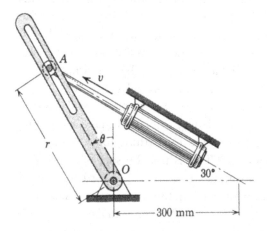

Figure 6

To do a correct reading of the problem the student must have mastered a vocabulary, not just of words as they are strung out in the statement but of symbols and markings as they appear in the figure. Here the shaded circle shown at the intersection of the horizontal and vertical lines at O is a *frictionless pin* (the slotted link is free to rotate about point O). The *heavy shading* below this pin and above the cylinder indicate *ground*; that is, the cylinder and the pivot point are fixed in space. Another *frictionless pin* is shown at A; it is attached to the rod emanating from the cylinder and is free to slide in the slot (it is not a threaded fastener). There are also letters and symbols representing variables and constant parameters—r, v, θ—and these must be distinguished from A and O, which identify points. Other symbols shown include arrowheads, some indicating the direction of variables, others simply the two endpoints of a distance measurement. Some lines in the figure show the outlines of physical objects, while others, usually lighter or broken, show direction or orientation or indicate dimensions as in a draftsman's drawing. A circular arc, with an adjacent notation of 30°, indicates an angular measure.

The student will have learned from class that the dot over the r in the text means the rate of change of the length r with respect to time. Two dots indicate the rate of change of the rate of change, and so on.

The marking "m/s" is a less esoteric abbreviation; it is to be read as "meters per second."

To the uninitiated, the figure might be taken as a design for an artifact. It is not. It is but a shadow of a design, not a practical representation that would allow someone in a shop to build a prototype. (This is not quite true without further qualification. If the shop were in the practice of making these widgets, and this figure was meant to show a variation on a previous model, then the shop could indeed make one. But if the only information available about the object is what is contained in this one figure—that is, if we ignore all contextual knowledge—then it does not suffice as a design.) So the figure is deceptive. Without a mass of background knowledge, you could make something that would look like this, but it wouldn't do anything for very long.

What does this thing do anyway? This is not the sort of question a student is supposed to ask. The thing does what the word text says it does, and nothing more is relevant. Indeed, the student is not supposed to give a moment's thought to how he or she might make one. This is not a design; it is a problem in the science of engineering mechanics.

The problem represents and reflects the analytical contents of the chapter in which I found it: "The Kinematics of Particles." In addition to mastery of a vocabulary of symbols and conventions of the figure/text, the student is meant to know something about syntax and the rudiments of creative conversation. More than the ability to read a drafter's drawing, the student must be able to see in the words and figure of the problem statement the workings of the conceptual entities and the mathematical relationships that tie the concepts together. The student must look through this shadow on the wall of the cave and uncover the "underlying form"—the guiding principles that constitute the formal and abstract structure of kinematics of particles. (Where is the particle [or particles]? We shall see.)

How, then, is one to do a correct reading, to move from a first confrontation to a successful doing of the problem, to find the answer? One might be tempted to start with the so-called hint found in the parenthetical remark:

Hint: Recognize that all acceleration components are zero when the velocity is constant.

As an attentive student, I would most likely find the value of such a blatant statement of a fundamental principle of kinematics to be questionable. "If the velocity is unchanging, of course the acceleration is zero. Why tell me something I already know?" Why, indeed. If I follow this line of inquiry, it might lead me to an inspection of the text. "Where is the velocity constant?" But there is still too much lacking; the connections are too diffuse, though we see the phrase "constant velocity" in the word text and the "v" on the figure.

I return to the top and read the problem from beginning to end, carefully, slowly, critically, chunk at a time, stubbornly refusing to move on until I have appropriated the text, both word and figure, to my own world of mechanisms and the kinematics of particles. I want to ferret out its underlying form, transform and reconstruct the problem statement so that I "get it," that is, grasp its analytical essence and then, only then, make use of the mathematical instruments developed in the chapter to determine "the values of . . ." Now, from the top:

> The hydraulic cylinder gives pin *A* a constant velocity of 2 m/s along its axis . . .

I see the hydraulic cylinder. It is the tube tilted up at 30° to the horizontal and fixed in space. (*Fixed in space* we already talked about. *Horizontal* is simply a convenient assumption—that the bottom of the page and the line parallel to it are horizontal. Does this matter? I don't know. But it's the kind of assumption that usually works, and besides, if it were important, I would expect some symbolic gesture in the figure to alert me to this "all other things being equal" kind of assumption not holding.)

But why "gives"? (Note the actors in this mythic tale. The "hydraulic cylinder" gives "pin *A*.") Cause and effect are implied. I can restate this sentence as:

> The hydraulic cylinder causes pin *A* to move with a constant velocity . . .

But how does a cylinder cause anything to happen, much less move with a constant velocity? It is the piston rod within the hydraulic cylinder that moves. So I will restate the sentence as:

> The piston within the hydraulic cylinder moves at a constant velocity and moves pin *A* at a constant velocity of 2 m/s along its axis . . .

That's still a bit crude and awkward. Better yet:

> The piston within the hydraulic cylinder moves at a constant velocity. The piston rod communicates this motion to pin *A*, which con-

sequently moves at a constant velocity, $v = 2$ m/s, along the axis of
the cylinder . . .

That's better. But then I could ask what causes the piston to move. What
are the forces (causes) internal to the cylinder acting on the piston? I
look but see nothing that might inform me on this matter. No hydraulic
lines are indicated in the figure, and no symbolic information about
measures of pressure are given. Evidently all of this is irrelevant.

At this point I am stuck and also frustrated. Here I went to all this
trouble, thinking hard about hydraulic cylinders, pistons, pressures, and
piston rods moving at constant velocity, and it is taking me nowhere. I
have wasted my time, gotten off on the wrong track.

But how is the student to know for sure that all of this is irrelevant
(the lingering wish that the effort wasn't wasted, that something might
still come of it)? How am I, in my creative reading and construction of
the problem, to know when to stop loading the text with supplemental
artifacts? Surely, some must be brought into the picture. Why is the
problem not about the power required to move pin A at constant
velocity, or about what hydraulic fluid pressure is required? How do I
know that this line of thinking is not going to be relevant to the analysis
at some stage, though not explicitly called for in the text?

The student doesn't know. But when you are stuck, you eventually
might allow that you've been following a wrong path. In fact, all of this
talk about pistons and hydraulic cylinder is irrelevant. I will do away with
it all and restate this first chunk as:

Pin A moves with a constant velocity $v = 2$ m/s in the direction
shown . . .

I jettison the whole business of cylinder, etc. I must amend the statement
to read "in the direction shown" since I have eliminated the axis of the
cylinder. My problem statement would now appear as follows:

Pin A moves with a constant velocity $v = 2$ m/s in the direction
shown and in turn causes the slotted arm to rotate about O. Deter-
mine the values of $\dot{r}$, $\ddot{r}$, and $\dot{\theta}$ for the instant when $\theta = 30°$. (*Hint:*
Recognize that all acceleration components are zero when the ve-
locity is constant.)

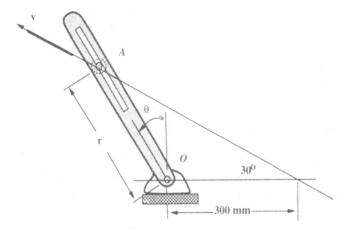

Figure 7

Now this nitpicking fabrication of false readings of the problem may strike some readers as nothing but a fantasy and a waste of time. A devil's advocate could claim that we all know what is meant by "The cylinder gives pin A . . ." But if we all know, if everyone knows so much, why do so many students fail in their attempts to do the problem?

Students fail because they misread and misappropriate the text. And reading is all that most students have recourse to, along with lecture notes and mountains of other similar problems. A few may actually have seen or felt a hydraulic cylinder or witnessed a mechanism of this kind in action, but how would this help? Indeed, this real-world, hands-on experience might even detract from a correct reading and prove dysfunctional by leading the student further astray.[15] Again, this is not a design; the essence of the problem is not to be found in a practical realization. The text, both graphics and words, is best seen, then, as a fantasy of sorts, a mythical tale, and the student's task is to fantasize in the proper way in order to construct a correct reading.

As in language learning, doing a good formulation requires the student to draw upon his or her recollections of similar formulations portrayed in lectures, the vocabulary of prior problems, and the voice and character of the technical problem-solving discourse that has pervaded all the student's other courses. All of this is what guides, constrains, and sets the stage for the encounter with this particular problem.

I continue:

Pin A moves with a constant velocity $v = 2$ m/s in the direction shown for an interval of motion . . .

What does "for an interval of motion" mean? Is motion to be considered as made up of intervals? If so, this is peculiar. Could I replace it with "motion's interval"? No, that won't do. Perhaps the author means an interval of distance or space or time. The latter is more likely. Although either choice will do, I will choose time. So we have:

> Pin A moves with a constant velocity $v = 2$ m/s in the direction shown for an interval of time . . .

Better, but I'm still uncomfortable. What's the duration of this interval of time? Clearly the pin can't move with constant velocity in the direction shown forever! The slot in the link only extends so far. If the author had provided another two dimensions in the figure, the length of the slot and the location of either end relative to point O, the maximum time interval could be calculated. (The length r is fixed by other given values.) I might estimate these lengths using a scale.

Hold it! I've wandered onto another false road. Another presumption in the baggage of normal engineering, science-based learning that students must accommodate is that, in a well-posed problem, *all critical information required to solve the problem, and only that information, is given.* Learning this axiom, found in no textbook, is essential to a student's survival and development. Problems that require "extraneous information" are considered to be dirty tricks.

But we have already eliminated some "extraneous information"—the description of the hydraulic cylinder. Apparently this kind of excess is legitimate to include, whereas the inclusion of numerical parameters that are not needed, or redundant, violates the norm. In the making of a problem, in the masking of the underlying form, only certain kinds of flamboyance are allowed (just as Aristotle allowed for spectacle even though he deemed such show unnecessary).

On the other hand, not having a value prescribed for this interval of motion, or rather this time interval, suggests that it is unsubstantial. I will therefore leave it out. Without a number, it means little. We now have:

> Pin A moves with a constant velocity $v = 2$ m/s in the direction shown and in turn causes the slotted arm to rotate about O . . .

That's clear enough. Or is it? One can see the slotted arm rotating counterclockwise about pin O as pin A moves in the direction shown. I see that pin A is free to slide in the slot. Still, some minor editorial embellishment might help:

Pin *A*, free to slide in the slot, moves with constant velocity $v = 2$ m/s in the direction shown, causing the slotted arm to rotate counterclockwise about *O* . . .

Now on to the next chunk:

Determine the values of $\dot{r}$, $\ddot{r}$, $\dot{\theta}$ . . .

Now *r* is a position vector of pin *A*. It is absolutely essential that I understand that *r* is not the distance to a point halfway up, in the middle of the slot, or some such point on the arm. To eliminate this ambiguity, I will remove the slotted arm from the figure, although I will leave pin *A* and the position vector *r*. I will also simplify the word text by eliminating all reference to this link. Thus:

Pin *A* moves with a constant velocity 2 m/s in the direction shown. Determine the values of $\dot{r}$, $\ddot{r}$, $\dot{\theta}$ . . .

Now the final chunk:

. . . for the instant when $\theta = 30°$.

Well, again, I'm confused. Is "instant" an interval of time, as in "brief instant." For example, does the author mean "for the brief instant when $\theta = 30°$"? *The* brief instant? Why not *a* brief instant? That's better.

. . . for a brief instant when $\theta = 30°$.

But as before, how long an instant? Does it matter? Are the values we hope to determine going to be constant, unchanging with time? Maybe this is why the hint is there. (I'm on another wrong track.) So it doesn't matter what length of time I consider to make up this "brief instant."

Alternatively, "instant" here may mean not an interval of time but instant in the sense of "the moment when," that is, a particular point in chronological time. So I try:

Determine . . . for the moment in time when $\theta = 30°$ . . .

This is better; it resolves the ambiguity in the word *instant*.

But how can I have something constantly moving at a moment in time? At each moment in time, it can be at only one point in space. We see here in the author's use of the word *instant* the vestigial remains of centuries of thought about motion, continuity, and velocity defined as a limit at an instant. Aristotle, Zeno, Galileo, and others may twitch in their graves, but the student ought not to get bogged down in useless "philosophical" diversions (a dirty word in engineering discourse). *Instant* means at a point in time; an instant has no "breadth."

Can we leave it out altogether, I muse. And why not?

Determine . . . when θ = 30° . . .

That works well. I am well acquainted with the one-to-one correspondence of distance and time as measures of position for a particle in a plane. Again, this is part of the baggage of the course. I posit a one-to-one correspondence of values of the angle θ and moments in time. The latter become superfluous. Chronological time doesn't matter at all. It could be yesterday, tomorrow, a century ago, a century hence. Whenever θ = 30, my solution is valid.

With this and one final change from "pin" to "point," my final transformation of the problem statement reads:

Point *A* moves with constant velocity 2 m/s in the direction shown. Determine the values of $\dot{r}$, $\ddot{r}$, $\ddot{\theta}$ when θ = 30°. (*Hint:* Recognize that all acceleration components are zero when the velocity is constant.)

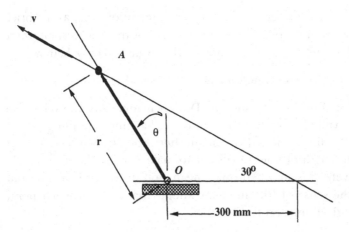

Figure 8

What have I done? I have transformed the problem into a mathematical exercise amenable to the analytical technique of the kinematics of particles—the relationships of position, velocity, and acceleration of a point—as found in the meat of the chapter. I have peeled away the stuff of the world and come close to the bone of underlying form.

The particle is the point *A*, *r* is the position vector locating the particle, and *v* is the velocity of the particle. I still have to choose a coordinate scheme to reference the scalar components of acceleration, velocity, and position, but these options are relatively well defined, and their evaluation is straightforward. (The origin of a plane, polar coordinate frame ought to be located at *O*.) And there is still some hard thinking required to keep straight which variables are given and which are changing and to be determined. It is not simply a matter of finding the appropriate equations in chapter text and "plugging in."

This is where the hint is useful as a sort of confidence check. As I work through the analysis, I will come to a point where I make use of the fact that, for a particle moving at constant velocity, all acceleration components are zero even if the particle is moving at constant velocity for but an instant![16] As I do this, I may recall the hint. Eureka! I know I'm on the right track now and why the hint was included, although it was a distraction up until this point.

The transformation I have made reveals the underlying form of the exercise. It is a "vector differential calculus" problem—abstract, universal, and unencumbered. There is nothing left of the mechanism save its essence (for this instant when $\theta = 30°$); no longer any pretense of machinery, hydraulic cylinders, piston rods, slotted arms, or frictionless pins. All of that is irrelevant. The answer need not make any reference to them; only the numerical values of $\dot{r}$, $\ddot{r}$, and $\ddot{\theta}$ were requested.

This, then, is the lesson of the problem set: The student must learn to perceive the world of mechanisms and machinery as embodying mathematical and physical principle alone, must in effect learn to *not* see what is there but irrelevant—all the ornamentation I have stripped away. This is why the figure is not a design; it is but a pretense of a mechanism, a fantasy construed to mask a problem in vector differential calculus.

This does not mean that the excess is pedagogically irrelevant. Indeed, it is absolutely necessary because it is what justifies the existence of a course in engineering mechanics. If all that were presented in the course were problems reduced to their core, this would be just another applied mathematics course. This masking—this fantasy of a mechanism hiding the motion of a point in a plane—is the essence of the construction of problems throughout all engineering curricula.

Thus engineering education in this respect teaches the student *not* to see. Students must learn to see through phony pistons and cylinders, the dimensions of an air tank, the ivy on the wall of Galileo's cantilever. They should go right to the heart of the problem, where lies the velocity as the rate of change of a position vector, the force that causes a rate of change of momentum, the angular lever of Galileo's cantilever, the linear spring of the truss member, the potential well of the silicon diode, the coefficient of thermal conductivity of the wall, and so on.

Reductionism is the lesson. Strip away the irrelevant, leaving only the small number of measures of physical things like position and velocity of a single particle—things that are related through the abstract and universal language of mathematics.

This is why Galileo stands on this side of the scientific revolution. Although, at first glance, the differences between his cantilever beam and the minimalist representations of beams found in today's engineering texts seem striking, both Galileo and the author of today's text are addressing the same object. Galileo's cantilever is deceptive to the modern eye schooled in technique. His attention to aesthetics—the ivy, the shadows, the grain in the wood—distracts. But for Galileo and his (small) audience, this late Renaissance image was a model of the same abstract sort as is used today. He, too, saw a cantilever loaded at its free end and bound to fracture at A, its fixed end. He too, saw a levering action—the weight E applied to the end of the lever arm BC and held in equilibrium by the internal forces acting over the lever arm BA.

Galileo's figure is hard to read as "scientific" because he wrote for an audience that was not accustomed to this kind of reduction, at least when discoursing on the mundane, everyday world about them. The reduction of the motion of the heavenly bodies amid the perfection of the celestial realm to mathematics was legitimate, but to treat the ordinary, corrupt matter of the sublunary realm in this way was rare. This is why Galileo's going to the Arsenal is significant—because he read and wrote about the ordinary and mundane, the beams, cables, and structures he saw there, in this reductionist mode. That is what makes him a revolutionary.

Now one might ask why the modern author of a mechanics texts, in loading the problem with all this useless and irrelevant information, stops where he does. That is, why is not more shown of the irrelevant setting? For example, one might further embellish the problem, some would claim make it more mundane and realistic, in the following way:[17]

> The linkage shown in the figure is essential to the function of a
> unit of semiautomated production machinery. [Include a new
> figure with two views, one of the machine and an exploded view,
> emanating from some point in the machine, that is the original
> problem figure.] Every four seconds the hydraulic cylinder is actu-

ated and the slotted arm swings into action, moving a protective
screen between the worker's hand and the workpiece. The hydraulic
cylinder gives pin A a constant velocity $v = 2$ m/s along its axis . . .

Of course this is going to take a bit more space on the page. Or how
about:

The linkage shown is an integral part of a piece of semiautomated
production machinery. Every three seconds the hydraulic cylinder
is actuated and the slotted arm swings into action, moving a protective screen between the worker's hand and the workpiece. The
workers are complaining that the pace is too quick and that the
screen often catches their fingers. The hydraulic cylinder gives pin
A a constant velocity $v = 2$ m/s along its axis . . .

Or while we are in this mode:

The linkage shown is an integral part of a piece of semiautomated
production machinery of U.S. Global Corporation's operations in
Papua, New Guinea. Since their buy out of General Radio, corporate directives have put a heavy emphasis on increasing productivity. Every two seconds the hydraulic cylinder is actuated and the
slotted arm swings into action, moving a protective screen between
the worker's hand and the workpiece. The workers are complaining that the pace is too quick and that the screen often catches
their fingers. They are threatening to strike. The hydraulic cylinder
gives pin A a constant velocity $v = 2$ m/s along its axis . . .

Clearly there is no limit to the creative fabrication of problems in
engineering mechanics, the manufacture of real-world problems in the
doing of which the student would learn to see the underlying form and
to disregard everything else. It is no wonder that engineers claim that
their work is value free.

The workers are no more irrelevant to the problem than is the
hydraulic cylinder. You say the slotted arm is there; but so is the worker's
hand!

But why is there this limit on acceptable kinds of mythmaking? (The
structure is the same across disciplines in engineering.) It is a matter of
values intimately wrapped up with the technique of everyday engineering practice.

Reductionism is pervasive; indeed, it is the essence of technique
within object worlds. There boundaries are drawn tightly, and the soli-

tary efforts of a participant in design toiling in the right problem-solving mode yield rich returns. Here, within object worlds, the payoff is clear.

But engineering practice requires a further reach. In the design of a mechanism for production, serious consideration must be given to the design of the hydraulic cylinder, the specification of fasteners, the unit's power requirements in operation—all of the features that were irrelevant to solving the textbook problem. An engineer working out the kinematics of particle motion of pin *A* (for more than just the instant when θ = 30°) is eventually going to have to deal with these matters (although it is unlikely that the worker will be taken into account as anything more than the source of a hand, that is, an *ergonomic object* or *user interface*).

Most engineering practitioners know that designing is not simply a matter of synthesizing solutions to independent problem sets. Although few of the complexities of engineering design process show up in the undergraduate classroom (save perhaps in an upperclass "capstone" design course), the working world of engineers is filled with negotiations across specialties, with decision making under uncertainty within contexts in which scientific principle is mixed in with social, political, and financial "constraints." The significance of the engineering textbook exercise and the object-world understandings inculcated in its mastery lies not just in that it undergirds all effective object-world work but that design participants take it as the basis for their attempts to deal with these complexities of process. How they do this, what they construct, and whether their formulations work or not is our next topic.

Process as Object

Participants in design do more than wrestle with problems within their respective object worlds. The overall design process requires that all of the individuals involved join together to plan, decide, critique, and integrate their efforts. In their attempts to make rational what experience shows is a highly uncertain and ambiguous concatenation of events, participants project onto this process of coming together the ways of thinking and working that they derive from their own object worlds. We turn now to some examples of object-world thinking that are embodied in instrumental methods whose purpose is to aid participants as they go about their collective work. Two characteristics of object-world thinking figure largely in these examples: (1) the urge to control complex affairs through a division and segmenting of process into independent components, and (2) reading time as either a resource or a metronome ticking away in the background.

Figure 9 shows the design process as it is depicted in many texts. This sort of map, intended for students of engineering design, shows how object-world thinking finds expression when its focus is on the design process itself. Typically, it lays out the process as a block diagram, with a progression through discrete stages—fifteen in this case. At the same time, it allows for backstepping or "iterating" around each phase (the arrowheads point up as well as down). The notion of different stages is common to most authors' renditions, some of which can be very elaborate.[18]

Such abstract figures express an ideal—an object-world creation of engineering faculty. Their intent is to establish control over the design process by breaking it down into discrete elements or subtasks, sharply bounding these subtasks by enclosing them in boxes or circles and then connecting them sequentially with straight lines. But while figure 9 and its kin may be useful pedagogically, in keeping with the reductionist tenor of such tools, as models of practical design activity they are deficient. If we allow the figure to direct our thinking about the people engaged in all the tasks contained in the boxes, we might conclude that design practice is an extremely orderly, rational process in which creative thought can be contained in a single box that yields a conceptual design or designs, which after detailed evaluation and analysis within some more boxes can be given real substance, tested, put into production, and then marketed for profit and the benefit of all humankind.

The diagram suggests a halting flow, a chaining of cause and effect; it might even be viewed as a conveyor belt, a machine through which the design is moved and acted upon, transformed and embellished at each stop. The only suggestion of possible messiness comes in the looping of some of the lines around the blocks. This indicates *feedback* and makes designing an *iterative process*.

A prerequisite to talk about feedback or iteration is the temporal ordering of the segmented stages of design. This entails definition of a clear beginning and end—the top and bottom of the figure in this case. The object as design process is then closed and bounded. Time, though not explicitly shown, is implied; it starts at the top and extends downward. We might even assume that each block ought to be allotted an equivalent amount of time. The orderly segmenting of process, with the design progressing down (falling) through this linear sequence of stages, suggests a form of determinism. True, allowance is made for a

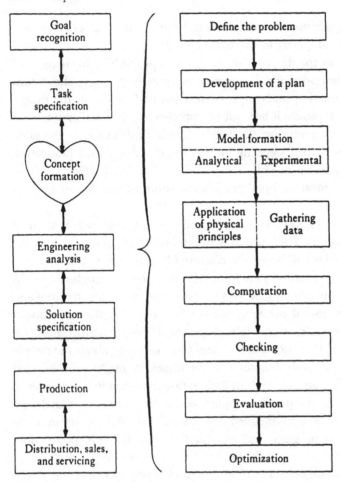

Figure 9

return to any stage after the completion of further stages in the process, but this is not readily apparent: The stuff within the boxes and their linear display simply overwhelm the thin lines indicating the possibility of feedback. Indeed, the ideal design process goes straight through; the form of the diagram suggests that iteration entails a cost, a cost in time. There is a "Tayloristic" quality to these visions of process,[19] though Frederick Taylor himself, I suspect, would have found them seriously deficient.

To anyone interested in process, these diagrams shed very little light on how design acts are actually carried out or on who is responsible for each of the tasks within the various boxes. Nor is it apparent what these participants need know, what resources they must bring to their task, and, most important, how they must work with others. The lines with arrows hardly represent the negotiation and exchange that go on within designing.

As a reductionist, mythical, object-world representation, figure 9 might be useful in the indoctrination of students into the ways of thinking of the world of the firm, but it misses the uncertainty and ambiguity of what really goes on in designing. Unlike the kinematics of particles, designing is not lawlike or deterministic. It is not a process of nature, nor can it be made to mimic nature. One can design an orrery, but an orrery that designs would not be designing!

There is a subtle aspect to work within object worlds that is not indicated in the block diagram: When your field of view includes all the participants in the design process, each of whom works within a different world though on the same project, each with unique responsibilities and interests, you must allow that, although they work on the same design, each sees the design in a way that conforms to the structure and language of his or her own object world. This multifaceted, or multi-masked, quality of design is not reflected in figure 9, which is an object-world construction in itself.

Turn back now to figure 5. Like the block diagram of the design process, the interim order processing flowchart can be seen as describing a method for controlling process, in this case the order/shipment process at Solaray. A close look reveals a sequence of numbered stages in a flow process. Without the numbers, the flow would be hard to recognize. But while the continuity of flow appears as secondary, like the implied hierarchy, the causal ordering of events is an essential dimension of this chart as it is to all object-world thinking. We see how (1) an order is communicated to Marketing Services from the order originator; (2) Marketing Services prepares a sales order and communicates it to Inventory Control; then (3) a work order is prepared (note passive voice); and so on. Note the lines and arrows indicating feedback of information to keep all interested parties informed of progress in the workings of this machine.

Information in two forms is processed: pieces of paper and verbal communications. The modules themselves, the actual hardware, make their way from inventory to the customer somehow, but that is not explicitly shown in the diagram. "Sales order," "production plan," and "invoice" all refer to the hardware, but they are like surrogates, abstractions and conventions that become media of exchange. They stand in for the artifact.

All of the forms are conserved. The pieces of paper get filled out, marked up, sent along, and filed away. They have a metric. They are numbered, and they account for things—modules, costs, charges. They are transmitted and flow along the solid lines from box to box. Loss of one of these artifacts, if it finds its way outside the channels depicted by the lines, is a minor tragedy. Note how the less solid substance of a verbal communication gets to travel down a leaky, dotted line.

The figure uniquely identifies all the object-world operations made on the forms. Not all moves are spelled out in detail. Only the major variables are seen in the foreground. As in any scientific modeling, the abstraction excludes as well as includes. By suggesting a single order in process, the design reinforces an object-world desire for clarity and certainty.

Each step takes up or consumes a chunk of time. The expenditure of time is only suggested in the diagram, yet the main intent in its construction is to ensure that delays in shipping or in meeting an order are kept to a minimum. Time is a resource to be saved.

This flow of events occurs in time, but not historical time. It is a block diagram of a process that is infinitely repeatable. An order enters; the process begins. The order is complete; the process ends. Another order enters; the process begins. But note the freedom from calendar time. Two, three, or even one hundred orders can overlap. We can have a block diagram for each order, if we like. The diagram stands free to be used again and again, independent of chronological time.

The figure conveys an illusion of simplicity. In its reductive, abstract focus on a single order being processed, it mimics a medieval representation of the earth orbiting the sun—a few Ptolemaic epicycles will do. If the epicycles are replicated, the motions of all the wandering stars can be explained and brought under control. So, too, our diagram is meant to control and manage the processing of many orders. It shows the same Ptolemaic confidence in a wandering star's inevitable return at a predictable time, ready to go around again, regardless of the epicyclic complexity or how many other stars it crosses in conjunction. Reduction to a mechanism of one controlling agency is its essence in accounting for many processes occurring simultaneously.

This chart is itself an artifact of a design process. Is it a good design? Yes and no. It "works," but not as a surface reading would suggest. We first note some of the inadequacies that are not evident in the diagram and then conjecture its true value.

The uncertainty of setting priorities at any stage in the order/shipment process is not shown. None of the complexity of deciding whose order to process first, whose next, is indicated. The figure has no depth. From the diagram one might conclude that the whole process could be programmed and transformed into a deterministic algorithm to be used across the board to deal with all orders and customers. There is no hint of ambiguity.

The design discounts the significance or stature of any individual, again in accord with object-world thinking. It is the person's function, narrowly defined for the specific purposes of this system of operations, that matters, not the individual named (or, rather, not named). The figure directs attention to the services and tasks to be accomplished within the boxes and away from the individuals responsible for carrying them out. As in the block diagram of design process found in the engineering text, the focus is on the artifact—the order forms and work orders as they are filled in, transmitted, and filed away. Who performs the various functions, much less what they do otherwise, is irrelevant. They are just so many hands required to perform a collection of mechanical tasks.[20]

In these ways, this production of members of a particular subculture displays the characteristics of object-world thinking. But while the reduction of a complex maze of bureaucratic affairs to a simple block diagram is powerfully pragmatic in setting a standard for getting things done, like any abstraction, something is always left out, and possibly something of importance not seen.

Consider an example. A customer wants information about shipping containers and the protection of the modules in transit. The order needs special treatment; the customer's request complicates the process. Along with final assembly of the modules from inventory and the receipt of a letter of credit, still another factor has to be considered. Who has responsibility for providing this information? The block diagram doesn't say anything about shipping containers. Different readings are possible. The first time through, the customer's request is tabled, put on the back burner within one of the boxes. It surfaces at a group meeting called to review the status of orders and inventory, and it occasions a heated discourse. Someone grudgingly takes responsibility for follow-up.

From an object-world perspective, such irrational behavior ought to be eliminated. This glitch in the system could be remedied by adding a new slot to the order form to account for special requests about shipping containers. One might go further and reconstruct figure 5, showing a box under one of the existing boxes in order to locate the new task within the process. But manifold other conditions remain;

indeed, the more complex the design, the more possibilities for things "going wrong."

Figure 9 is an academic production; figure 5, a production of engineering practice. Within the engineering firm, other maps and methods are made and employed in an attempt to control the design process. A Milestone Chart, for example, provides a snapshot of a block of time in the life of a design participant (or project). It shows how that time—a period of several weeks, a month, or even several months—is to be "spent." Tasks are listed in a column at the left; time, measured in increments of days, or weeks, goes from left to right across the top of the chart. Even though time's arrow is indicated and time as chronological time implied, the calendar dates of the start and end of the period of the chart are left open. That is, the maker of the chart fills these in. The clean chart initially shows only blocks or spaces whose horizontal lengths represent time-as-resource, and that resource, as duration, comes in finite chunks.

There is a similarity between the block diagram of the design process and the milestone chart in this respect: In the block diagram, time's arrow is implied, going from top to bottom. And the blocks can be read as each taking up a finite chunk of time, of being roughly of the same duration. Yet the events depicted stand free of any point in chronological time. Indeed, the exploded view within figure 5 makes this clear; the vertical position on the diagram is not to be construed as a particular point in time. The block diagram, as the empty form of a Milestone Chart, is meant to stand free of time, ahistorical, true for all time.

Time, then, is a resource to be used, saved, or wasted, but time also serves as a reference. Time taken in the latter sense is clear in the explanation a senior executive gives of problems in coordinating different activities within the firm—the development of an innovative machine tool for the production of a photovoltaic module, the design of the module itself, and the introduction of a new technique in cell processing:

We have a problem of getting things to move in sync. . . . We have real problems if things get out of phase.

This is not time as a resource. In this statement, there is no indication of lack or shortage of time. Rather, the image is of developments evolving according to their own inner logic as time goes by at a steady pace—like a dynamic system with more than one degree of freedom, each likely to gyrate autonomously if a senior executive doesn't exercise his power and authority to keep things in phase. This is the language of control theory (the object again as icon).

The statement presumes a common measure, ticking away in the background, providing an independent reference. The different activities may move more slowly at times, speed up at other times, but they must all come together at some common point in time if they are to be productively employed.

Another technique used in the management of design process, intended to keep control over diverse but interrelated activities, is the Critical Path Method (CPM)—a method of scheduling whose "output" consists of a graphical display of the sequences of steps that must be followed, and how they must intersect and come together in time, to ensure that a design deadline is met. Like the text's image of the design process, the CPM chart focuses on the artifact, the thing being designed. The graphic image shows key events (as circles perhaps), telling when supplies arrive, when they are integrated into the design, when the design is frozen, when a subsystem is to be tested, when marketing gets into the act, and so on. Looking at the network of interconnected circles, you see the artifact coming together—like the image conveyed by running the film of an accidental shattering of a vase backward.

Film is an appropriate metaphor: Object-world time flows by uniformly and continuously, as it does in a theater. But a film editor in the studio can stop the flow at any point, work with chunks of time as a *resource* to dramatic effect. So, too, the design practitioner, technician as well as executive, attempts to control the time available as a resource. On the other hand, for the theater patron, time is hardly noticeable unless its continuity is broken by uneven dramatic construction—it is time as *reference*.

From a Cartesian perspective, there are not two kinds of time. Even as I write to illustrate one or the other, I find it difficult to keep the two distinct. I have pointed out how time in both forms, as resource and as reference, is present in the block diagram of design process and in the Milestone Chart. Similarly, in the CPM chart, time as reference flows in the background, but the user must construct estimates of duration of each event—time as resource—before the method can be put to use.

Both conceptions of time reflected in these methods reveal its antiseptic character within object worlds. Time in science (except to those who wonder about the origins of things) and in engineering is either duration or a rootless, endless, fixed metric against which one can value duration. Time as historical moment plays no role within object worlds or in these methods intended for the management of design process.

As Henry Ford would have it, time as chronology, as history, is for the most part absent from object-world thinking.

Milestone Charts, CPM charts, block diagrams of order/shipment processes—all are artifacts and productions of the subculture of the firm. The way in which they figure time conflicts with the sense of time felt when we observe design process. As long as we speak of time as resource alone, a static, synchronic view of design process suffices. But when we turn to time as reference, to time going by in the background, it is difficult not to face the drama of events and, as observer, try to fix this unfolding in chronological time. As observer, I cannot easily use time as a reference for events without locating those events in real, chronological time. We face here a unique diachronic phenomenon.

The events I observe continually prompt the question, "What happens next?" There, observing, I am in the theater wanting the next reel to begin without a hitch from the last, a fade-out/fade-in, not to be noticed. Participants in design are also patrons in a theater, not in the editing room. They cannot really turn the film backward, edit, or rework to make the design process go smoothly as much as they would like, as much as their reading of time as referent, stripped of historical contingency, suggests.

It is not surprising, then, that the methods designed to manage and coordinate the design process as it moves through time are deficient in the same way that any method that looks to the future is deficient. They can never be sure of events to be, since no one can predict the future. This uncertainty is what gives designers an opportunity to develop creative, cost-effective, robust, and innovative designs. The design process is about the new and the uncertain by its very nature. If the process is sure from the start, you can be sure that it is not a design process.

If these instrumental techniques intended for the control of process are as deficient as these examples suggest, how do participants manage the design process? Why do they make and use these artifacts if they are so full of holes? To answer these kinds of questions we must get inside the boxes of the figures, not to draw more boxes but to see how participants do what they do. As an example, we return to figure 5 to find out how this design of the order/shipment process was itself constructed.

The meeting starts with the usual joking, guffawing, and teasing. No one, not even the division manager, is immune from assault. Hierarchy appears to be totally lacking; the group appears to function as one harmonious social unit within which barriers to exchange and restraints on allowable discourse are minimal.

The meeting goes smoothly for the first hour. There is the customary review of Milestone Charts and the status of the marketing and design projects that currently claim different individuals' attention. The engineers in the room, Ed and Tom, describe the delays they have encountered in getting the module support structure defined for the new, large, residential module. Brad, head of Marketing and Systems Engineering, makes some announcements about shifts in the responsibilities of marketing members that appear not to be news to the affected parties.

A full hour has gone by when John, from the marketing side, raises a question concerning the way in which customer orders are handled:

Can the originator of an order be informed when an order goes out?

There is an edge to his request; it urges an explanation of the current process and implies that it isn't working. In part, the query can be understood as a straightforward question by a relatively recent new member of the division, but there is a crispness to his question that suggests a complaint as well as a query.

Mary, his colleague in Marketing Services whose business it is to oversee the routing of order forms, responds:

We are working on the problem. You know how the order is supposed to move. You enter the order, it goes to marketing then on to panel fab, inventory, . . . shipping.

Evidently, Ed from Engineering also has a complaint:

That's too late for the information to get back to me. Can't you send a copy of the shipping order to the originator?

Tom, his colleague, enters the fray, beginning to describe ways they could fix up the process:

Why can't Jane serve as a central place for information about the status of an order? If it went to Jane, that would fix it.

Ed muses about the possibility of having the computer play a data management role.

In this exchange, a sense of frustration over the amount of time it takes to fill an order is evident. That soon surfaces explicitly.

Mary begins to offer an explanation of how a shipping date is specified:

Ten days after the order date . . .

This raises more antagonistic speculation. John presses for a further explanation. Why, after all, ten days?

It appears that there is to be no easy resolution of what now all agree is a "problem." Ed proposes that they try out some "scenarios" that might help them understand the process in toto. He wants to understand how the system is supposed to function and how it does function, here and now. By drawing out a scenario in the life of an order from start to end, from customer inquiry to final shipment, he hopes to grasp the workings of the order/shipment process in its full generality.

But Mary wants to describe how a shipping date is set, and she brings Paula, who handles Inventory, into the picture:

Paula . . . assigns the shipping date . . . and then five copies.

Ed and Tom, almost in one voice:

Five copies. And where do they go?

Ed, before anyone can answer, gets up from his seat and goes to the front of the room, to the board. He wants to map out the process with a block diagram. He asks:

What happens when the shipping date is assigned? Five copies . . .

Mary tries to continue to explain Paula's role, but Ed wants to move away from Paula and talk about the whole system and how the order makes its way through the process. He raises another issue:

Who then has responsibility after the date is assigned? I'm trying to understand how the system operates.

John and Mary have difficulty joining in this kind of analysis. John persists in trying to patch the system on the spot, trying to get some agreement about change. He notes:

It would be bad if each one of us calls Paula . . . better if Jane checks.

Mary says she sees that as her responsibility, but John points out that she is often away from the firm, on the road, a fact Mary acknowledges.

Meanwhile (this recent exchange has taken but twenty seconds), Ed wants to play out his scenarios. The impasse reached in John and Mary's tête-à-tête gives him the opportunity to retake control of the floor. Ed continues:

That's what we are talking about—we need a central person. But first let's get the kinks out of the system.

And he sketches a scenario on the board in the form of a few blocks. The attention of the group, other than Tom, is not given readily. They wonder whether this kind of extensive analysis is necessary. Ed continues his block diagram:

The driving force is the box marked "Mary."

John replies that the driving force is the person who types the form.

The sketch develops into a maze of intersecting lines; boxes multiply as all the actors and their functions are identified. Ed, again:

. . . have to streamline . . .

Tom joins Ed at the board. John and Mary carry on independently as Tom erases "Mary" from the critical box and replaces it with "shipping clerk." "Jane" is likewise changed to "secretary." Tom is explicit:

Let's get the system working independently of the names.

John suggests that the problem doesn't warrant all this abstraction:

. . . don't think the job is as large . . .

Ed and Tom contest this strongly. Ed replies:

A good flow diagram ought to be generated.

With Ed and Tom's insistence and a nod from Brad, they agree to continue with this analysis of the problem, but not at the current meeting. The session draws to a close with John urging all interested parties to continue participation in this process. All assure him that they will do so.

The object-world reading of figure 5 contrasts with the process whereby it was first given form at the meeting and subsequently re-worked by Ed into the form you see here. Its true value and usefulness to the group are also not apparent from the figure alone.

Throughout the meeting, participants remained sensitive to the in-terests of each other, although the discussion at times became heated and personal. The flow of ideas was turbulent. Participants often re-peated or returned to claims and questions they had voiced earlier. Coalitions seemed to form but these were often broken and reshaped as new questions arose. While there had been some discussion among participants prior to the meeting—John had talked with Brad, express-ing his intent to bring up the subject at a group meeting—they really weren't sure where they were going. They didn't have a sheet of perfor-mance specifications.

Amid this disorder and uncertainty, meanings could be misconstrued, interpretations truncated or extended beyond the speaker's intent. Words like *order* are fuzzy in this context; one person may see this as a document processed internally within the firm, while another will see the customer and his or her requirements; that is, the order is both a form and a customer. So, too, an operation on an order is not simply a filling in of a line on a form but a particular participant's responsibility. So, too, for Ed and Tom a block diagram is an efficient and complete representation of process, while for Mary and John it is a sidestepping of the real issue. While the figure appears relatively crisp and clear, the discussion that was its source was not.

Constructing the problem—laying bare what was wrong with the order/shipment process—was akin to laying out a set of performance specifications for how the process ought to work. It consumed most of the participants' energy. Various scenarios had to be exercised (exorcised?) to make things clear. All of this consumed a kind of time later referred to as "wasted." No conservation of time as resource here.

Participation in this process demanded a balance of critical comment and suspension of judgment, a willingness to listen, and at the same time a desire to make precise, to define, to fix and order thoughts. In this process a high tolerance for ambiguity is to be valued. Indeed, in any design process, as people work on their concepts and images, specifications and constraints, there must be ample room for shifting images, rethinking concepts, changing specifications and constraints.

The content and direction of the meeting were dependent upon the actions of individuals—that John, not Ed, brought the issue to the fore; that Mary, not John, had been working on the problem. At times it appears that chance events and personal proclivities governed the process. Ed's going to the board and sketching a scenario was decisive in turning the discussion toward the production of figure 5. It was this act of inscription at a blackboard that cast the discussion into the object-world form of a block diagram of process. From an object-world perspective, what was until that moment a confused array of personal interests, historical anecdote, ambiguous expression, and uncertain goals now became clear, or at least a framework for making it all clear was set. The session culminated in agreement that (1) there was a problem, and (2) the analysis ought to be continued in the direction Ed had laid out—with a block diagram of process.

This analysis points to a thicker reading and interpretation of such charts and diagrams. They should be seen not only in object-world terms but also as artifacts of agreement negotiated by participants. Despite their appearance as descriptions of the workings of a system—marked by hierarchy, boundedness, clarity of form, and a determined and causal character—their real meaning and true value are to be found in their making. They are significant primarily to those who gave shape to the symbols and forms that represent their often discordant interests and responsibilities. As such, like a political constitution, they often require a new reading in the wake of their construction.

A Lighter Side—Process as Object

We have seen how the characteristics of object-world thinking pervade the artifacts produced by participants in engineering design, extending beyond the realm of hardware and the modeling of physical systems to encompass the planning and management of the design process itself. Before we move on, I want to note some other artifacts as evidence of the projection of object-world thinking onto the world of process.

Tacked on some wall in every engineering firm is a statement, or list of statements, commonly labeled Murphy's Law: "If something can go wrong, it will." Variations on this dogma abound. It is a joke. You read it and are supposed to laugh. Yet it is more than a joke; it carries a message about process that is truer than a block diagram. Things are always going wrong in designing, or appear to do so, when events are compared to the expectations embodied in a designer's story making, Milestone Chart, projection of development costs, and so on.

Why do things always go wrong? Is it intrinsic in the second law of thermodynamics (that things are by nature degenerate)? Is it an object-world reality?

Design, by its very nature, is an uncertain and creative process. In every design task there is an opportunity for creative work, for venturing into the unknown with a variation untried before, and for challenging a constraint or assumption, pushing to see if it really matters. Uncertainty in one sense allows participants to exercise their creativity. But uncertainty in another sense ensures that there will always be unforeseen outcomes. Call these the unknowable. There will always be the unknowable, and this is the root of Murphy's Law. It simply expresses recognition of the contingencies of creative work.

Of course, after the unknowable happens, we reconstruct our vision, and the unknowable happening becomes a possible but perhaps uncertain outcome. There still remains the unknowable. This is the sense (proof?) of Murphy's Law: There always exists the unknowable; and there is a high probability that the unknowable will be "wrong" when it becomes known; hence, if something can go wrong, it will. The phrase "if something" is the ironic bit because there is always something; to say "if something" is as much as to say "something will go wrong."

Other expressions of the true character of designing as process are revealed in drawings of things impossible. What at first glance looks like a useful fixturing device or tool turns out to be, on closer inspection, a trick drawing made to deceive. Again the absurd is chosen to indicate the possibility of deception by the hard, yet ambiguous, world of the artifact. This time the symbol points to the formal drawing intended for manufacturing and the possibility of surprise even here. For example, the structural members of the frame for the new residential photovoltaic module are shown in three views as required by the manufacturer. Beth checks Tom's work and asks how he is going to get the leads through to the junction box. There is no way; Tom has forgotten to leave sufficient clearance. His drawing now joins the ranks of the impossible.

Still other expressions of uncertainty are heard in phrases like "not invented here" or "idiot-proof." The latter suggests an imperative of design—the product ought to be made such that even an idiot can work it. So participants in design who abide by this norm work to shield the innards of their machinery from the user. They multiply internal redundancies, make it well nigh impossible for the user to get inside and poke around, then add "idiot lights" to inform and direct the operator to call the service department if something goes wrong. Setting impermeable interfaces with potential users is one way in which engineers try to extend their control over the functioning of their productions out into the marketplace. If the unknown happens out there, there is little they can do to make it known. Not all subcultures abide by this design norm. Still, it is part of the lingo and, again, albeit absurd, expresses recognition of the uncertainty in the life of design.[21]

And then there are stories in the form of fables about life in the firm before the change in management, about the trials and tribulations of getting the desalination plant on line, about Product Design always coming in at the last minute and asking Technology Development to fix its problems. Some of these become company lore, passed around, held in common trust, trucked out on occasion for the enlightenment of the new hire. Others are more functional and have direct technical import.[22]

You laugh at these stories. You are supposed to laugh. The absurd is not to be taken seriously. Yet seen against the framework of object-world thinking and object-world artifacts, they can be seriously seen as an expression of a design culture's realization that object is not enough— that in designing, order, hierarchy, cause and effect, and so on are not enough. Room has to be made for disorder, chance, the creative, surprise—in short, the unknown.

Perhaps we have spent too much time focusing on object worlds. It is, as I have said, only a part of the design participant's day, often a minor part (but still a significant part). It is the proper place to start, however, because the ways of thinking and seeing within object worlds provide the framework for the totality of every design participant's efforts.

It is time now to move away from the object, its iconic influence in designing, and its management to look at the context of design in another sense—as ecology. We have already made the point that the furniture of object worlds is more than the rules and concepts of scientific theories, because it includes all sorts of supporting networks of resources. We turn now to explore, on the level of the firm, the infrastructure of the subculture and the design team.

5

Ecology

It is Monday and I am sitting in Michael's office, listening while he gives me an update on progress with Amxray's cargo inspection prototype.

It's getting close to our deadline on systems integration of the A-to-D with the detectors. Al is coming in this Thursday and we would like to be able to show him we're ready to hook up.

Al is BG's project manager. He comes to town roughly once a month for milestone events such as a formal design review or, as in this instance, a systems integration test. I relate to Michael how I had seen fistfuls of cabling strung out in Jim's bay downstairs and running down the main corridor. He explains:

That's what it takes to hook up 512 detectors to the A-to-D unit, and keeping order is no small job. But Jim has a good system. And remember that cable has to be long enough to reach well outside the test chamber—he ought to be out of the hallway in a day or so. We're shooting for Friday to hook up. We will make it in good shape if that damn connector tool shows up.

I ask:

I thought you were going to make one in-house?

He responds:

We thought about that, then decided we didn't have the time or the people. So we're getting the job in Nevada to free up their tool. That took some talk. You ever see what we're talking about?

I reply that I had not so he goes on to explain:

It's a small thing, fits in your hand. Got to have it to extract any faulty connections, and we screwed up on a few. We had the thing here, then they needed it out West. Nevada didn't want to let it go . . . claimed they used it every week now . . . wanted to know why we couldn't get another from Connecticut. Jeez, they were stubborn, and here it was ours in the first place. I had to get the boss to set them straight.

I begin to wonder about this special tool that might very well become the straw that breaks the camel's back. Why was it so special? Why couldn't they make one in-house that would meet their immediate needs?

I knew that assembling leads together, keeping them in the correct order, and fixing them securely within the sockets of a connector is a chore, one that is almost impossible without the appropriate tool. Indeed, it is an awkward task even with the right tool in hand. And if a mistake is made in locking a single pin in place, as revealed in testing the electrical continuity of the line from one connector to the other at the opposite end of the cable, still another special tool is needed to force the pin free of the incorrect socket. It was this tool that was unavailable to Jim and was threatening to delay completion of the assembly of the 512 leads to connectors prerequisite to the first trials of the x-ray inspection system as an integrated whole.

Later, I called the firm that produced the tool and obtained a drawing of the device. I was surprised by its simplicity. It is hand-held and functions much like a hypodermic needle. The whole instrument is no more than five inches long. A black, phenolic ball at one end nests in the palm of the hand. At the other end is a "probe." To extract a pin connector, you insert the probe over the mating end of the socket, depressing a clip that catches and fixes the pin in the socket. When you squeeze the tool between your palm and forefingers, a plunger moves down the center of the probe and forces the pin free of the socket.

The probe is made from steel drill-rod stock and has a black oxide finish. Its tip is held to tight tolerances and hardened to the level "Rockwell C-45-55." The probe sits inside the end of an aluminum cylindrical sleeve. The plunger slides along the centerline of the sleeve, through the probe. A compression spring acts to return the plunger to its initial position after it has freed the pin. Spring and plunger are made of stainless steel, but of two different stock varieties. There is a retaining ring that both guides the "action" and holds the probe within the sleeve. The plunger's diameter varies along its length; it is least at the end that slides within the probe; at its other end it is threaded to mate with the phenolic ball.

That's all there is to it: ball, plunger, probe, sleeve, retaining ring, and compression spring. There are a few peripheral ingredients: a label is stamped on the body of the sleeve identifying the model; there is a protective plastic cap encapsulating the probe that must be removed before use; and no doubt a sheet of printed instructions accompanies the packing, which consists of a carton and some styrofoam pellets.

The simplicity of this element of Amxray's infrastructure is deceptive. Its ingredients are themselves dependent upon a broad range of materials and machines and processes. Phenolic, stainless steel, anodized aluminum, the bit of plastic encapsulating the probe, even the styrofoam packing chips may not present the full range of modern materials but do indicate the variety of what is available and essential in engineering work. Think, too, of the different metal-cutting and forming machines that turn down the stainless steel plunger, bore out the hole through the probe, drill and tap the phenolic ball, cut the threads on the mating end of the plunger, and coil the steel spring out of wire stock. All of these are ingredients of today's infrastructure.

These materials and machine tools depend in their turn upon still other materials, machine tools, and processes for their own existence. We can envision a web of interconnections among the elements of infrastructure, a sort of CPM chart of all the antecedents to what Jim had called the ninety-nine-cent connector extraction tool.

This is not a static display. If you sit and observe the web over time—a few months will suffice—you will see change. New varieties of materials appear, and the cost of production processes changes. Massive machine tools are increasingly computer controlled, and so our web of infrastructural elements reaches out and connects with that of another industrial sector—electronics.

And if we attend to the spatial locations of the sources of materials and machines, we could not fail to notice the national and increasingly global reach of our web. While the machining of the sleeve and plunger might have been done in a job shop next door in Connecticut, or even in-house, and the retaining ring is available from any number of suppliers, the phenolic ball might be an off-the-shelf item ordered from a supplier on the West Coast, and the compression spring might be made from Swedish steel and coiled to the requisite dimensions out in Cleveland.

Yet even this inventory is inadequate. Thinking solely about materials and machines yields a limited, object-world rendition of infrastructure. If we take off the blinders, we must contend with other dimensions.

Envision, in addition to this web of strands and lines leading from the connector extraction tool back through all of its parts, materials, and machinery, another intersecting web showing financial transactions, communications and information-processing protocols, letters of credit, import duties—all the machinery of national and international commerce. Still another network includes skilled workers, machinists, and technicians.[1] Closer up, I see designers at the extraction tool firm itself figuring out what spring stiffness is required and how much space to allow for its working. Lines go from them to reference books and catalogues, lines from there to historical precedents and textbooks of engineering mechanics. I see now an educational web intersecting with manufacturing and commerce, with nodes representing engineering methods for figuring springs, for drawing and computing—modern computational techniques that allow designers to estimate the spring's behavior before it is ever made real. I am lost in a maze, a complex, multidimensional web of interconnections that is the infrastructure of contemporary design. (Do you know how your telephone works?)

All of this is contingent, more fragile than it first appears. The firm situates itself with respect to a particular constellation of resources. This network shifts and reconnects as innovation provokes change within the field. The principles of a spring may not change with time, but the new hire may have a whole new bag of computer tricks with which to model their performance in a particular device. Throw out the old; bring in the new. Or a supplier goes bankrupt—another change. A new process for heat treating, or an altogether new material, or an alternate source of a needed component or process appears on the scene, and new connections are made with expectations of a better design, a lower cost, and a broader market.

Infrastructure expands at one place but contracts at another. "Sorry, we don't stock that any more. We were just getting too few calls for that odd size spring." There is no historical law that says it must evolve or even that it must change slowly, though historically the time for change in some of its dimensions might be measured in decades. And continuity is not assured. There need not be another supplier who still produces the "odd size" spring.

But designing is itself dynamic and contingent. It is the new that is designed. If the infrastructure can no longer supply the odd size spring, there are alternative designs, perhaps more robust (perhaps more costly), that call for a standard size, although some may think that the "feel" of the "action" in using the tool may not be as good.

The infrastructure of design is seen in this initial cut as a domain or field of resources—of materials, machinery, processes, tools, instruments, and enabling networks of supply, information, finance, and distribution. Resources appear, too, as shared visions of the possible and acceptable dreams of the innovative, as techniques, knowledge, know-how, and the institutions for learning these things. Infrastructure in these terms is a dense, interwoven fabric that is, at the same time, dynamic, thoroughly ecological, even fragile.

Field of Constraints

The infrastructure of design is also limiting. Not all dreams are realizable. Knowing and judging when a move is innovative or folly are figured within some field of constraints. Infrastructural constraints include scientific precepts, all manner of rules, and explicit and implicit norms.[2] We turn to these now.

The character of specifications and constraints varies widely. Some are formal, in the sense that they are written out, set down in a memo, a contract, a purchase order, a reference text in the firm's library, and the like. Others are informal, part of folk knowledge, only explicitly articulated when challenged.

Some derive from scientific theory and practice. For example, in the design of a photovoltaic module, the open-circuit voltage of each and every cell lies near 0.5 volt, the voltages and currents in an electrical circuit obey Kirchoff's laws, and the resistance of a conductor is a function of the wire diameter and length; these are all "facts" and "laws" that constrain. They rule over designers' practice and serve as a basis for the story making within object worlds, enabling designers to model, to quantify and work up their specifications, drawings, and orders. Yet scientific constraints do not by themselves define the artifact. Despite their apparent pervasiveness, they hold only within distinct, tightly circumscribed object worlds.

Other kinds of technical constraints are more clearly human constructs. The interface conditions among different subsystems, worked out by the project leader in consultation with the key people on the design team, are one example. The performance requirements of a customer are another. Although explicitly articulated at the outset, these constraints should not be considered inflexible; because they are constructs, they are subject to change. Still other constraints, including regulations promulgated by government agencies or codes published by groups such as the American Society for Testing Materials or the Institute for Electrical and Electronic Engineering are negotiable. Like common law, or any other human construct, there is always a need for a reading and interpretation in their application and always more than one way to meet their intent. Far from being ex cathedra commandments, these human constructs derive their meaning out of their continual exercise and redesign. One need only look at the amendments to codes or witness the negotiation of specifications with a customer or client to see how lively dealing with this sort of constraint can be.

Scientific concepts and laws must also be interpreted in their application, but any "negotiation" here is within an object world and tightly constrained by the value-free presumptions of the language of science. Readings and interpretations still must be made; but whereas a code or performance specification is like a norm that a design must meet—think of falling short or exceeding a prescribed "safe level"—we do not speak of meeting scientific laws partway. One doesn't ring up and appeal to a higher authority to ask for a change in Kirchoff's law to accommodate one's special, unforeseen circumstances.

This is not to say that technical constraints in the form of regulations, codes, or interface requirements are often challenged or easily changed. They—excepting government regulations—do not appear to cause much complaint because they have come out of the work of the design and engineering professions themselves. Developed over the years en famille, they become part of habitual ways of thought and action, a dog-eared page in the code book. They derive from the experience of professional specialties represented within the firm and within the design culture at large.

Government regulations differ in that they are the productions of a broader world of more diverse interests. From within the object worlds of participants in design, congressional subcommittee members and the witnesses at their hearings appear as foreigners; most legislators as technically illiterate lawyers. A constraint of this sort is seen as "not invented here," and its intrinsic object-world content is viewed with suspicion if not disrespect by participants in design. That is why an externally proclaimed regulation prompts complaint and why good, workable, acceptable yet effective regulations are hard to write and implement. The problem is not simply in the drafting; it lies as much in getting the regulation accepted as legitimate and relevant to work within object worlds. The process of constructing and implementing a regulation works best when it is, like the design process itself, one of negotiation and exchange across the myriad of constituent interests. This is no light task.

When time and the resources to proceed in this way are not at hand and interests are so much in conflict that they cannot be bridged, yet those with the power to legislate (or amend existing rules) claim that the need is clear, a new regulation can still be effective. Its promulgation and implementation in this case can lead to abrupt changes in design practice and a firm's ways of doing business.

Regulations, in the sense described above, were infrequently encountered in the design of photovoltaic modules at Solaray or the inspection system at Amxray, or in the solution of the dropout problem at Photoquik. Constraints on design in the form of codes were much more common. We will look at the constraints engaged by Solaray in designing the residential module's frame and support structure prior to field testing.

A major New England electric utility had expressed interest in testing photovoltaic technology as an alternate source within their system. They wanted to see for themselves how much electrical energy a photovoltaic array on a residential roof would produce in order to estimate how it might affect their distribution and transmission system. They were well aware that this technology was still far from cost-effective when compared to conventional power sources, but they saw it as being potentially important for the future, especially if its costs were to come down as its advocates prophesied.

Solaray's new residential module appealed to them. They liked the appearance of the company's line of smaller, low-power modules and were assured that the larger residential module, when its design was completed and production begun, would look even better. They contracted for twenty-five residential units, each system to be rated at 2 kilowatts, with an architect who had had considerable experience with residential photovoltaic systems. He, in turn, subcontracted with Solaray for the modules.

Ed, one of the systems engineers at Solaray, is assigned to work with the architect to develop a framing and support structure for the modules to "interface" with the roof. The product of this effort will have use beyond this one installation. The architect welcomes Ed's participation.

Creating an interface between the roof of a house in central Massachusetts and a photovoltaic module is a mating of "low-tech" and "high-tech," a matching of vernacular craft, that of roof rafters and joists, shingles and flashing, hammer and nail, with the science-based technology of photovoltaic cells encapsulated in heat-treated glass with modern thermoplastic backings and framed in aluminum. Now there are codes that prescribe limits on the amount of weight a roof structure can support. And there are ways architects and contractors have of estimating the distributed weight due to winds and snow loading on roofs of various pitch. In New England, several feet of snow can accumulate on a rooftop. Tests have been run, and their results are also available to Ed. The codes and data provide a starting point for his design.

Ed proposes to support the modules with rails running along the joists up and down the pitch of the roof. He argues that this will distribute and transfer the load evenly to the supporting roof structure. The architect doesn't like the idea. He's got the job of installing the modules, and he sees Ed's proposal as labor intensive. He proposes supporting a framed module on four short legs or feet. These "standoffs" are much easier to install.

Ed, in his turn, sees this as more work for him and more expense for Solaray, and rightly so. He will have to take responsibility for the design of a frame that will be rigid enough to support the module by itself alone. His original proposal had the rails acting as both frame and support. The differences would not be major for this one job, but they could entrain significant increased costs further along in the res-module design effort.[3]

Ed questions the four-legged design. He is concerned that with all the weight concentrated at four points on the roof, the loads might exceed what the code specified as allowable. However, in his search through the codes he is not able to find a statement that answers his specific question. The texts and codes he reviews prescribe limits for distributed, but not for concentrated, loading.

Ed marshals his resources. He reviews studies of wind loadings on buildings with solar panels on their roofs, explores the consequences of science-based rules of static equilibrium for his particular situation, tests the four-leg design against the available building codes that specify maximum allowable distributed load. Making some assumptions and using his tacit knowledge of the local strength and flexibility of roofing and its ability to distribute a concentrated load, he sketches out different designs for the feet. He conjectures how they could be fastened and sealed in order to better distribute the load into the joist. He develops a sense of how much quantitative allowance there seems to be in the codes themselves—the object world of vernacular New England roof structure clearly differs from his aerospace world.

In this way Ed constructs a story—an analysis that predicts what loads the joists will see under what he considers to be severe weather conditions of high winds coupled with snow loading. The exercise leads him to believe that the four-point support is not bad after all; indeed, it is feasible. His story ends with:

Under conditions . . . and with feet of this design, we should be within code when we translate into a distributed loading. I've tried to take a worst case estimate of local loading and what that would do to the joist and roofing locally, but if we make the feet this way, and seal in this way, then we should be OK.

All agree to go ahead with the design as conceived by the architect and then modified and elaborated by Ed.

Codes are constraining, but constraints guide as much as they restrain. Ed uses the codes in an indirect way. Because they speak only of distributed loading, they do not apply directly, yet they do provide him with a framework for analysis and some crude estimates of the magnitudes of allowable loads for his situation. He reads and interprets the existing code, extending its reach to touch upon the problem he faces.

This is not unusual in design. Codes, in the business of designing the new, often have to be given a fresh reading and a new interpretation. A code is a historical statement based on experience and testing, but mostly on experience. Design is design of the new and the untried, the unexperienced, the ahistorical. As such, the code can never assure the successful and safe performance of the artifact or system in all of its ways.

Craftspeople as well as engineers learn from error, from the unknown making itself known in a dramatic, sometimes disastrous way.[4] Failure is a common source of codes. That is why they change with time. History is always being rewritten in this business.

Ed renders a reading of existing code, extrapolates in object-world terms, and then justifies that extrapolation using all the resources available to him. His story goes unrecorded. The results of his effort are reflected in two kinds of formal design drawings: in the detailed drawing he makes of the feet and in the assembly drawing the architect makes of the full array on the roof. But the story he constructed is gone. His assumptions, presumptions, tacit feelings, hard-nosed calculations, and confident voice are gone. The code has been applied, guiding as much as restraining the design of the structural support of the rooftop photovoltaic array.

Yet it may very well be that Ed's work becomes one item in the further development of codes. It is after all a historical fact that twenty-five photovoltaic arrays do now sit atop as many single-family residences in Massachusetts, and to date the roofs have not leaked, caved in, or been blown away. They are there for the future development of code—say a code for the support of photovoltaic modules on rooftops in New England. In this way a code is a human construction.

There are other constraints that are clearly negotiable in the more typical use of the phrase. Cost constraints, for example, which at first sight appear to be as hard and precise as scientific law, share the same characteristic as codes. They are aspects of a social consensus whose true reading depends on a local knowledge of context.

In process, when drawings are sketches and concepts as well as details of "the way to go" are changing day to day, costs, too, are in flux; a supplier's quotation or a project leader's estimates are soft; there is ample room for give-and-take. It's not just that reevaluation of the cost of some item or feature is always possible, but rather that the item itself may be redefined and reshaped into something not quite the same as the old—again the unanticipated, the unknown. It's not that the shooter misses the mark so much as that the target may be shifting.

Costs, then, are not simply absolute numbers tacked on to a proposal, a parts list, or a contract at some well-defined stage in the design process. Rather, they enter into design deliberations in the same way as a performance specification, an interface requirement, a code, or the like. In discussions among participants, cost estimates appear as negotiable, changing as the more physical characteristics of the design change. Sessions at which costs are invoked are reminiscent of a poker game; estimates are bid to justify one's own proposal or to critique and call another's.

A conversation heard in the hallways of Amxray:

Mark *ABC Graphics has a new board out that will do most of what we want.*

Michael *Will it improve our timing? Will it give us a smoother transition from shot to screen?*

Mark *It's bound to. It looks like their hard-wired logic will work very efficiently in our application, and at only $300 a board.*

Michael *Does it have enough on-board flexibility? Or are we going to have to do a lot of engineering to interface with it? How much of an effort there . . . and what other stuff do we have to add to our system . . . power supply? Have you added all that in?*

The cost estimate intermingles with response time, with added engineering at the interface, and with the need for other hardware. Mark states a cost of the board, which to him looks reasonable. Michael counters with an implied larger cost estimate of what it will take to get the board working, integrated with the rest of the system. At this stage in the design process these estimates need only be very crude. They themselves are constructions and malleable—remember that one can move the target. Beyond the significant figures are other measures of the cost of an option, such as previous positive experience with a supplier or, to the contrary, unease with a quotation from a firm that always seems to ask for a two-week delay in shipment.

Scientific laws, cost estimates and budgets, performance goals and interface specifications, IEEE codes and ASTM testing procedures, environmental regulations and the like are all part of the ecology of designing. Perhaps they ought not even be considered in the same category, since they are so varied—some deal with politicians, others with electrons. They also differ in that some lie in the background, as the dictates of science, while others appear explicitly in the foreground, as in a contract or a purchase order. Yet they are the same in the sense that they are interpreted, given a reading, if not constructed almost out of whole cloth on occasion—in short, they are negotiable like the characteristics of the design created anew.

They are also ecological. Participants in design both draw on constraints to sustain and guide their design efforts and feed back into the environment a redefinition of a cost, a particularized performance specification, a new reading of a code, even—sometimes—an extension of scientific theory. This new environment sustains the next phase, the next design effort.

A vignette further illustrates the ecological nature of constraints.

Early in the design of the x-ray inspection system, a meeting was held at Amxray to try to define the performance specifications of the full-scale system intended for the inspection of large cargo containers such as truck trailers. The design of the scaled-down prototype system, which would be able to scan containers the size of an aircraft cargo container, was at midpoint. The way and means for storing data had been written into the contract for the prototype (by disk)—and that design, which had focused on the performance of the hardware alone, was well on its way to closure. But specifications in this regard for the full-scale system had yet to be set.

Participants at the meeting included Arnie, the project scientist; Jim, from Electrical Systems; Michael, the project manager; Al, visiting for the day from BG; and a few others, in and out, over the course of the morning. There was discussion of different performance features of the full-sized system, including the following discourse on the archiving of data, of images:

Arnie   *I think we are going to need two operators running the show. Someone is going to have to be able to get the image up on the screen . . . not just that, but get it up in a form that shows the most—*

Al   *Why two? That's going to get expensive, two technicians.*

Arnie   *No, not two technicians. You need a tech to get the image on the monitor but we will need another more experienced viewer, a radiologist, to review the images that looked like there was something there.*

Michael   *What's that do to our throughput? What's it going to take to do a good reading?*

Arnie   *Got to have a disk for temporary storage so the radiologist can get a good look.*

Al   *Why not archive all images if we're going to have temporary storage?*

Arnie   *No, that wouldn't be practical . . . would take far too many tapes.*

Jim   *It's not just the volume of tapes. This also has implications for our software.*

Al   *We could have the software tailored to the customer.*

Michael   *It seems to me we are getting into a lot of philosophy. In terms of nuts and bolts, what's it going to cost to archive all this, to do what Al would like to do?*

Jim   *We might go to an optical disk, at 300 megabytes. But that's high risk.*

Michael   *We've got a possible export licensing problem there.*

Arnie   *Still think we ought to only archive those that are suspicious?*

Al   *Is it impossible to do what we said, archive all cargo containers, or can't we at least make room for that option?*

In this exchange, participants state a variety of concerns. We hear proposals for hardware and conjectures about how many people will be needed to run the system and what their professional capabilities should be. We hear estimates of the quantity of data to be archived, how it will be stored ("far too many tapes"), and a claim that all of this will have an important effect upon system software. At one point a reference to "a lot of philosophy" brings the discourse back to more narrow, strictly technical terms. (*Philosophy* has strong negative connotations in engineering discourse. Here it is used to turn the discussion.) There are questions about costs and even the legal implications of export regulations.

This spectrum of concerns entrains a variety of potential design specifications. It is a discourse in which nascent constraints are being formed and options set out, and it is hard to distinguish here between the two. Participants are creatively exploring options while, at the same time, defining, redefining, fabricating, and reconstructing constraints. Under the umbrella term *archiving of images*, we can envisage the team formulating specifications for operator training as well as for data-storage hardware, setting requirements on system software, and even developing variant designs—one for export, one for the domestic market.

These different options will be pursued in object-world exercises and sorted out and redefined in mostly informal meetings of the different participants. Not all of the concerns aired will be considered seriously. Some will be taken to the fore; others will be shunted aside as unworthy of attention. At this stage, at the end of this meeting, however, it is impossible to say for sure which concerns, ideas, or proposals will suffer what fate.

The process by which participants move from this unstructured exploratory state to the detailed specification of hardware, the making of drawings, and the drafting of documentation, training manuals, and contractual disclaimers depends upon the organization of the task for the next step. One important factor is who is given responsibility by the project manager, their way of seeing things, the importance they place on this phase of their efforts, and the resources they can bring to bear in sustaining their proposals. Different paths are possible.

Consider, for example, the question of the qualifications of the operator of the system. This concern may lead to the specification of a substantial training program based on the premise that the operator will take an active role in the interpretation of images as well as in the maintenance of the system. Or it may lead to the design of sophisticated software to ease the burden on the operator, making the task of discriminating contraband from normal cargoes as automatic as possible. In this case the operator is seen as just so many hands, idiot-proof becomes the governing design norm, and the training program becomes secondary.

Whatever design trajectory is followed, or rather constructed, is a matter of individual interests and values, of the organization and traditions of the subculture, as well as of the hardware and software, the costs, and the codes and regulations of data storage and image enhancement technology. The construction of constraints in the archiving of images is a social process: What are today's ideas and concepts in flux are tomorrow's constraints.

Organization

The web of suppliers, vendors, fabricators, transporters, educators, and financiers is one domain within which the firm situates its productive work. Without a proper siting and know-how about the most viable connections and opportunities for improvement, the firm's work would not be competitive. The field of constraints is another. Without a mature, working sensitivity to the implications of constraints, the firm's designs would be destined to fail. Turning inward, within the firm a more immediate infrastructure sustains participants' efforts. Organization, informal as well as formal, structures relationships among participants in design. It provides modes of communication and facilitates participants' negotiations; it provides physical resources (tools, work stations, fax machines) and legitimate charge numbers (procedures for accounting and keeping track); and it governs access to the infrastructure outside the firm's four walls and sets the stage for the culmination of design—the "launching" of a new product. Organization is people situated in a formal hierarchy, but it is also grapevines, taboos, and unwritten rules and norms.

Organization is construed here as a third dimension of the ecology of designing. Participants in design work within the formal organizational structure of the firm and take that as given. It was what they were hired into, and it therefore has an air of permanence that may in fact be unwarranted by the historical record. In its division of the firm into subunits by function, it is meant to support and control the workings of its members.

Organization charts come in different levels of detail showing different levels of the firm, like the boxes within boxes of the design-process diagram. They describe a hierarchy and try to indicate the different functions of individuals or groups of individuals and their relationships. The president appears at the apex of the most global image of the photovoltaic firm. Branches lead from this point to the leaders of five groups—Research, Production, Marketing and Systems Engineering, Comptroller and Administration, and Employee Relations. Within each of these five groups, there are further subdivisions. More treelike diagrams are drawn to show the finer details of the organizational state of the firm.

These diagrams also have a Cartesian character. They show a clear separation of individuals and groups and a clear distinction of functions linked in an apparently unambiguous topology. Yet this formal structure, though it provides a supporting network to the individual, is generally inadequate to sustain the work of design. Engineers and managers need a more flexible, ad hoc structure—one focused on the artifact rather than on the competencies of different members of the firm (e.g., accounting, marketing, engineering, manufacturing). In this the artifact of design itself provides a template, a rational way of organizing the efforts of individuals. I call to mind *Zen and the Art of Motorcycle Maintenance,* where Robert Pirsig, in an attempt to explain the workings of his machine, envisions breaking it down into parts or subsystems—ignition system, power transmission system, and so on.

This perceived hierarchical and functional structure of the artifact can and often does serve as a pattern for the organization of design.[5] Hierarchy is evident in the way in which the division of the design task into more manageable part is made. The project leader and the division management organize the design process to attend to different subtasks and levels of detail. The design of a large photovoltaic module is broken down into subsystems: the module glass and frame structure; the electrical networking of the cells, diodes, and connectors; the materials to be used in the substrate; the manufacturing processes; and the marketing plan. All of these subsystems and their interfaces are formally defined in a document.

In the design of the cargo x-ray inspection system, five different design tasks were identified and set out as constituents of the whole, namely: (1) the detectors, (2) the x-ray beam collimator, (3) the data acquisition unit, (4) the control and display unit, and (5) the system software. Each of these subsystems performs a different function: the collimator collimates the x-ray beam, the detectors detect, the data acquisition unit acquires the data and puts them into digital form, and the software massages the data for the control and display of images. One could break these five areas down further, for example, by splitting software design into two tasks: the design of systems-level code and the design of operations-level code. This process, in principle, might go on without end. Here, then, is another instance of how the object of design and object-world thinking can structure the organization of design and hence the relations among participants.

The reduction and naming of parts of the design and the pattern it yields are not unique. They are not inherent in the system or object itself. As in any classification scheme, one must choose the criteria by which one dismembers a particular domain and fix words that point to the pieces. A motorcycle or an x-ray machine or a crisis design problem can be "cut up" and "named" in more than one way. This cutting up is metaphoric; spatial and topological divisions will usually not serve as boundaries. A simple physical piece of the whole answers to more than one design criterion; it is the stuff of more than one object world. On the other side of the coin, a single name can point to more that one constellation of artifacts and their functions, depending on whom you talk to and where their responsibilities lie.

The nonuniqueness and ambiguity are reflected in the challenge of setting clear interface conditions between subsystems and hence between different design tasks. The division of labor pretends to efficiency in design as in production—another reflection of the hierarchical and reductive nature of object-world thinking. But in the design of motorcycles the power transmission group cannot design completely independently of the ignition system group; in the design of x-ray inspection machines, the data acquisition design must interface with the design of the control and display unit; and in the design of a large photovoltaic module, the design of the cell circuitry is coupled with the design of the enveloping structure through the junction box(es). Where you locate the boundaries between different design tasks is no straightforward process. Who ought to have responsibility for what element of the design? This question is intimately wrapped up and interwoven with another one: How are the different parts constructed, chosen, and named?

The artifact itself, the object of design, can only serve as an organizing principle in a fuzzy way; the fuzziness shows up at the edges, where participants from different object worlds must come into contact to discuss and negotiate their differences. Last year's model can serve as a source of names and a division of tasks, but the new design effort is likely, if it is a true design experience, to challenge the old divisions and stimulate a new vocabulary.

Faced with the organic and unknown of the object of design, one might conclude that any organizational effort intended to break the design task into independent subtasks is a gloss, a deception, even an attempt doomed to failure. A critic of instrumental ideologies might jump on that bandwagon. That would be selling matters short. A more constructive reading reveals that while the goal of independence can never be fully realized, some division of labor is absolutely necessary; it is how far one moves along that path that is an open question. What is essential is that participants recognize the continuing need for exchange and redefinition of boundaries across the boundaries themselves.

But there is something more going on here. This act of breaking up the total design task is an act of design itself. That is, to engage in this organizational effort requires participants to formulate concepts, to imply consequences, to name and label. Here is designing, too.

Just as performance specification and constraints are not simply given "boundary conditions" within which one designs, so the formal organization of the design effort is not an inflexible initial condition. Organization, as one dimension of the ecology of design, is part of designing itself. There is a give-and-take between the two.

In all of this, the biological metaphor works well. Design acts and events are embedded within nurturing fields—organization within the firm, infrastructure outside, and constraints throughout. This environment configures design; it shapes and sustains the substance and spirit of designing just as designing ultimately reconstructs the field.

Focusing still more inwardly, gazing for a moment on instances of "micro" organizational action, we reach a similar conclusion. The Critical Path Method and its relatives can be seen as attempts to gain control over resources external as well as internal to the firm—a supplier's delivery date, an external design review, budget approval for phase three by corporate headquarters. They help maintain ecological health and balance through continual exchange. Their application also strongly influences the design.

In the construction of a CPM chart, there is a necessary process of negotiation among all groups that are part of the network of resources, and that have an "interest" in the design process. When Tom at Solaray calls a supplier to get an estimate of a delivery date on the junction box for the new residential module, he is negotiating. Tom wants a reasonable estimate, one that fits his schedule but is not overly optimistic and thus will not be met. The supplier has to think of other jobs in the pipeline and rank their importance relative to this one. "If I work hard to meet his request, I know that I can do it, but will it really be rewarded in future orders? What information do I have from Tom that this new product is going to fly?"

That's just one circle on Tom's CPM chart. There are three other suppliers to contact, and then he has to make sure Steve in Panel Fabrication is tooled up for the new size glass. And when did he say we would have delivery on the glass? And so on. By the time he has entered all this information into his CPM software package, he has made promises, received commitments, and in the process reset his and his colleagues' thinking about the design of the module. In other words, the construction of the CPM chart is a design act: It contributes to the definition of the artifact, how it will be, how it will perform, as well as what it will cost.

We have seen how the subculture of any firm works within an infrastructure. The word *ecology* suggests an organic process; the firm, or better yet the design team, is situated in this infrastructure and is sustained by it while at the same time it contributes to the resourcefulness of the infrastructure. The word also suggests a process of exchange, not just in economic terms but in social terms as well. That is, the vitality of the design process, as revealed in the interactions of the design team both internally and with the rest of its world, depends upon the socially construed value participants place upon each other's work. Their transactions are best considered social, in the sense of Mauss's *The Gift*,[6] rather than utilitarian acts guided by economic self-interest. The efforts of participants carry value in exchange because they bind the expectations of one participant to another. Belief and trust in the value of the work of another are essential to the design process.

To illustrate the social and ecological nature of the organization/design nexus within the firm, we will describe Sergio's marshaling of resources to sustain an attack on the dropout problem. Sergio was a member of the Advanced Technology Division of Photoquik. Their formal, stated mission was to take a long-term view; to pick up promising ideas, novel and generic, and explore their potential for use down the line in the next generation of print products. Several of their developments had led to substantial design improvements. Many of their efforts led to naught, but that was not unanticipated. And besides, there was usually some knowledge gained even from negative experiences—they knew not to go down that particular road in the future. They defined their projects in consultation with people from all over the firm—from Servicing, Marketing, and Product Development—and they kept a critical eye on the competition's latest technological developments. For the most part, they set their own schedules, since they were focused on future rather than immediate results.

This was not the nature of Sergio's immediate task. He knew he would be working in a crisis mode; Atlas wanted his solutions yesterday to the problems they had encountered today. It was not the first time Advanced

Technology had been called upon to work in this way, and Sergio himself had previously been through this kind of exercise.

The formal organization chart, along with the formally stated mission of Advanced Technology, does not help much in times of crisis—let's call it normal crisis—such as this. The task was complicated by the fact that the Colorado division of Photoquik, though it had what appeared to be the same dropout problem in their new product, was not connected to Sergio's efforts in a direct way. For Sergio to obtain resources, in particular a budget allocation, he would have to convince Leonard in Colorado that this was a legitimate undertaking; but organizationally this was not a simple matter. A formal request for an allocation to work on a problem in a remote area theoretically had to go up though several levels of the organization and back down and around again for approval.

Yet Leonard was not remote in a social sense; he and Sergio had had many phone conversations and had visited each other's turf. Leonard's problem was real, and Sergio saw that it needed a solution.

Closer to home, the Atlas team, facing time constraints much more stringent than Leonard's, had also approached Sergio. They knew that they could not expect a complete, elegant solution to incorporate into their machine before full-scale production began, but they were hoping to learn from his efforts and, what's more, get a crude fix, something to carry them through until a better solution could be retrofit in the field. With this in mind, they had their own people working on a solenoid-activated cam. Sergio found that they were willing to support his effort by providing him with the hardware to test out the options he chose to pursue. He spoke to them about funding, including an outside consultant if necessary.

He had been told of a firm, QB Technology, that described itself as idea and development people. Sergio thought of contracting with them to evaluate some of the fourteen options independently, untainted by the prejudices of people within the firm. They might also develop new ideas, and he instructed them to feel free to do so. This effort would take a relatively small amount of funding, which he felt he could cover with allocations from Leonard. So, with verbal assurances of support, he had Contracts draw up a purchase order for QB's services.

In-house resource people were not easy to find and break loose. Hans had been with him from the start. Advanced Technology management had made good on their promise to provide him with at least one person possessing the requisite skills and knowledge of the photoprint process.

Hans and Sergio worked well together planning an approach and contacting individuals within the firm with different kinds of expertise who might be able to spend some time formulating options. Indeed, this is where the fourteen options had come from—from Sergio and Hans's meetings, informal and formal, with Chemistry people, others in Research, in Advanced Technology, and even in Production. Hans was particularly useful in getting Chemistry people to spend time on the problem; they were the recognized "authorities," closest to the "soul" of the firm's products, the main line of their patents, and still revered (and highly paid). Hans was a process chemist himself.

Don came on board not too long after the second of two concept generation and evaluation meetings. He would work on the air knife concept and then move to what looked to be one of the more promising options, the E&M excitation of the bed. Sergio negotiated for still another engineer; he wanted someone to serve as liaison with Atlas and also take on responsibility for a promising option that had been added late in the game—a redesign of the print-paper path to effect a greater contact force with the bed. It was an apparently low-cost option in the sense that making the change would not require a lot of engineering or alteration to the existing hardware. Atlas seemed to like the idea. The down side was that no one thought that this conceptually simple change would solve the problem 100 percent—maybe 50 percent, but not completely. Still, it was an option well worth pursuing, and he wanted someone to work on it full time. Advanced Technology management went along with his request, and Hector came on board the following week.

Sergio needed hardware. It wasn't too long before Hector and Don wanted a machine to explore possible mechanical configurations for the acoustic drive and paper-path guides, respectively. Atlas had promised machines, but they were slow in coming; and when they did deliver, it was with a caveat: they would need that particular machine back soon and would replace it with a newer version. That was OK with Sergio as long as they gave him two weeks' notice to plan for the change.

Getting technicians was at least as difficult as freeing up a few engineers. Phil was a contract employee, in and out of the plant on different occasions and jobs. Sergio would have preferred a more seasoned in-house technician, but Phil had worked on Atlas in its early configuration and was therefore was familiar with the hardware. That was a plus. Sergio was satisfied when Don reported positive things about Phil's ability.

With the machinery, the actual hardware, an individual responsible for each option, and technicians all available, serious work could begin. This was not the end of Sergio's negotiations, however. When he argued in favor of Don's suggestion that a drive roller be moved, Atlas would have none of it, despite the strong rationale Don offered. It was too late in the game for that kind of change. They did approve Hector's proposal for altering the guide path, though. Sergio didn't quite see the difference but was happy to extract at least some recognition that Atlas might have to change the configuration a bit if they wanted a robust rather than an ad hoc solution to the problem.

All of this was duly reported at the weekly group meetings. While Leonard's liaison at the firm was invited to these meetings, he rarely showed up. Atlas occasionally sent someone over, but more often Sergio was on the phone or in immediate contact with them. They were just a (long) hallway away.

By this time Sergio had laid out a plan, a schedule of development and testing of the three options—the E&M device, the guide-path change, and a change in the chemistry of pretreatment, an option that would require little in the way of mechanical alteration of the current product. Sergio liked having three quite different approaches; there was a value in their independence. Since they were uncoupled, in the sense that each of the three would work independently if installed in the product, they could be developed independently. Furthermore, each should contribute independently to the solution of the problem: If the guide-path change took care of 50 percent of the problem, then the E&M device could take care of the remaining 49 percent of the dropouts. And if that wasn't good enough, chemical pretreatment ought to take care of the rest. It was also possible, of course, that one of them would prove to be a solution by itself, even the guide-path change, but Sergio was marshaling his resources in parallel. The object of his tests was to uncover the critical parameters of each option, how they could be integrated in parallel, and what the effectiveness of each was under different operating conditions.

He contacted the Taguchi Test people to get them thinking about how he would best set up the testing. Then he could go to Leonard in Colorado and to Atlas closer to home and say, "See, here is our proposed fix, or set of fixes really. You buy the E&M device and you get this performance. Chemical pretreatment gives you this at this rough cost. Guide-path change gets you this far along, but it's a steal, and you ought to make that change anyway. What do you want to buy?" That, at least, was the plan.

Reflect now upon Sergio in the firm, working within a structured organization on an unstructured design task. In truth, most design tasks are just as unstructured at the outset. But the crisis form of Sergio's task makes this more evident.

Sergio must bring structure to the task. Drawing on the infrastructure of the firm, including his immediate subculture and divisions further afield, and resources outside the firm, he must marshal, construct, demand, and negotiate for what he needs to do the task—people, machinery, time, and funding. But he must do still more.

He must make his design task legitimate and respectable in the eyes of those he calls upon for resources, grants contracts for services to, orders parts from, and promotes his solutions to—all those who intersect and interact with him in his efforts to solve the dropout problem. He must fabricate a small world within the firm where he and his design team are given their due. He must put together that team and get them working on bits of the problem that challenge them—bits they can work on independently yet bits they must see as parts of a unified whole. There is just one problem—the dropout problem—and there is just one design team. This is necessary if Sergio's task is to be seen as legitimate and worthy of attention in the form of responses to phone calls, funding, machinery, travel money, outside consultants, respect for delivery dates by suppliers, and so on.

Indeed the problem itself must be constituted as part of this process. Formulation of the problem—by Sergio and Hans together with the specialists they call upon—is part and parcel of designing a fix to the problem. The vagaries surrounding different representations of dropout match the variety of the representations themselves. Sergio has to build consensus among all interested parties that will allow them to talk about the core problem in a common language. The problem has to be reified in a particular way, so that the meaning of the label *dropout* is clear and coherent to all. Sergio is designing a problem as much as a solution to the problem.

Sergio furnishes his space within the organization, an organization that formally has little room for this kind of crisis relief activity, with committed people, procedures and milestones, schedules and budgets, contracts and purchase orders, design options and test procedures, print machinery, guide-path mechanisms, E&M devices, and the design problem itself. He and others on the team must now see to it that others in the firm and outsiders recognize the significance of their work and so be responsive to their call. Sergio must make his ecological niche.

If we allow the object to fix our view of designing, we see only hard edges, formal drawings, irreversibly machined and mated pieces; we see invoices, parts counts, and costs figured to fractions of a penny; we see organizational pyramids and isolated subsystems; we see engineers working to meet a customer's needs that have been clearly expressed as a set of performance specifications; we see work on tasks and subtasks within a well-defined, given, fixed field of constraints. This view is static, ahistorical, and rigid—a lifeless landscape in which everything is rationally determined.

In fact, designing is otherwise. By focusing on the ecology of design, we have begun to see that participants in design work within a rich, multidimensional environment that reaches well beyond the narrow confines of their own object worlds. A customer's needs are not given or discovered but must be created; an operator's capabilities must be defined; building codes need interpretation; costs must be tried out; budget limits must be agreed to. The task must be organized into subtasks; suppliers must be coaxed to commit to a price and a delivery date; the dropout problem at Photoquik must be constructed. All of this is designing. In all of this, choices are being made, decisions foreshadowed, and possibilities discounted.

There is, I realize, a danger inherent in the use of metaphor in an analysis like this. I have introduced the word *ecology* because it allows me to stress the organic and hence dynamic, vital, and reciprocal character of what might otherwise be construed as a linear, deterministic process. I have argued for a perspective that sees the flux of design affairs as ecological in the sense that the infrastructure of the hard object at its center extends in many dimensions, is continually shifting, dissolving old boundaries, and creating new possibilities, and is even fragile. So, too, is Sergio's task of maintaining the design team in pursuit of a fix for dropout.

6

Design Discourse

Dropout

The first thing Sergio did when called upon to "put out the fire" on Atlas was to canvas the firm seeking advice and soliciting suggestions from individuals who had worked on the development of the photoprint process and its realization in hardware. In the course of this pilgrimage, he constructed a list of fourteen possible solutions to the dropout problem and then scheduled a meeting of the dozen contributors to his list in order to do a preliminary evaluation of the relative worth of the various options. He intended to apply the Pugh Method and narrow the number of options down to two, three, or four—some small number that he would shortly have the resources to pursue.

The Pugh Method, intended to aid designers in the early stages of the design process, prescribes a sequence of steps to follow in order to evaluate possible options against criteria established by members of the design team. The evaluation is to be done in concert, at a group meeting. Critical to the method is the initial setting out of an unambiguous statement of performance specifications—functional and other characteristics the final artifact must exhibit in order to be considered a successful design. These provide the basis for the articulation of criteria against which options are to be judged.

Think of a matrix to be filled in. Along the top, labeling the columns, are the different criteria. Along the left edge, running down the page, are names labeling the concepts—one row per concept. Each participant fills in the boxes of this matrix column by column, marking in each box their estimate of the option's potential to satisfy the column's criterion (relative to all the other options).

Before the team starts the comparison, they choose one concept as a baseline or reference configuration. The others are then rated against this usually less imaginative or more traditional option. The degree of fineness of the scale of measurement is arbitrary. The simplest scale might allow a concept to be Better, The Same, or Worse than the baseline.

A process of iteration is used to reconsider and improve upon the options. After the matrix has been filled once, the team reconsiders each option in turn and, focusing on those criteria that are poorly met, tries to reformulate the option to eliminate its deficiencies. The process goes on for cycle after cycle until the matrix shows a clear favorite.[1]

This is something different than science-based work within object worlds. The underlying form of this methodology for decision making is not described as a mathematical theory in a textbook. Instead, the technique is suggestive of the classification schemes of the taxonomist, the organizer of species, seeking niches and criteria to bring order to the world. It has the flavor of empirical Baconian methodology, in which one lists all the characteristics of an observed phenomenon and then stands back and infers a structure that positions the objects within a field and enables instrumental analysis.

After coffee and doughnuts, Sergio begins:

Sergio *OK. Let's start. You all got this.* [He holds up a description of Pugh methodology.] *I sent it around last Thursday. It pretty much says what we're going to try to do, except I'm going to make a few changes. You'll see as we go along. The basic idea today is that we want to first set up some criteria to judge. Then we compare how the fourteen go, compare them against these criteria. By the end of this morning I'd like to have narrowed things down, not to one option, but to three, say, something we can get going on. Yeah, Harold.*

Harold *It says in this method that we ought to pick a baseline option to compare against. How are we going to do that? It seems to me any one of the fourteen would be as good or bad, for that matter, as any of the others.*

Sergio *I thought about that, and here is what I propose. Let's pick the option we know best, OK? Say the QWP. We know how that works, and other than that it probably won't fit in the space we have to play with, it still can be our reference. But first we have to set up some criteria. So, let me get this chart around here.*

Hans *Obviously we need a criterion, something like "Gets the job done" or "Eliminates dropout."*

Sergio *Yeah, that's got to be one. The thing has got to work, to solve the problem. How did you state it?*

Marco *What do we mean when we go and claim that, say, the QWP eliminates the dropout? I mean, all of those up there have a chance of doing the job.*

Sergio *I know. But we score, not with numbers but say three, four marks—better than the baseline, say the QWP. This is where the baseline comes in. Second would be neutral—no better, no worse than the QWP—and third would be negative; that is, we think it won't be as good as what we know works now.*

Marco   *Yeah, but some of these options I think might work as good, even better on some papers but probably won't work at all on others. How do you grade it then?*

Sergio   *What do you mean? Give me a more specific example.*

Marco   *I mean like with the air knife. It might work with Z-weight paper, but with the heavier M-weight I don't think it will work.*

Hans   *Why not make that another criterion: "Works with all papers."*

Sergio   *Or "Sensitivity to paper." Sort of pull that out from under "Does the job."*

Marco   *You mean that there are some options that will do the job, but some of those won't be able to handle the heavy paper?*

Sergio   *Yeah, that's one way to look at it. "Does the job" is our best guess that the thing will work, but we give paper type a separate category. We may want to say something else has to be done to handle the heavy paper; that becomes another problem.*

Fritz   *How do we know whether paper type is critical for the air knife? It seems to me we don't really know what the problem is. How can we compare options when we don't know what is causing the problem?*

Marco   *Fritz, that's a good point. Do we really know enough to—*

Sergio   *We know we have dropout on Atlas. We know that the QWP gives good results. We have a pretty good idea of what consistency it takes to give good print—print that a trained eye can't find a hole in. (With a magnifying glass, you still see some.)*

Fritz   *Yes, but we can know, and should know, a lot more before we go judging these proposals on whether or not they will solve the problem. If this place hadn't cut back on its chemistry research, we might have a chance of knowing what the hell is going on, not just with Atlas but we had it on Mars as well.*

Sergio   *Look, some things are beyond our control. We have no power over the powers-that-be. We don't have a chemistry group working on this problem to call up and say "Get over here and help us evaluate these options." We've got to go with what we have. Atlas is due to go out onto the streets in seven months.*

Fritz   *That's the way it always goes around here. Someone wants your solutions yesterday.*

Sergio   *OK. So we have "Eliminates dropout" and "Sensitivity to paper." What are some others?*

Hans   *Cost.*

Marco   *Have you guys thought about some kind of chemical pretreatment . . . different papers?*

Sergio   *Cost. Let's think about that. Is cost really that important? Leonard says he doesn't see cost as really significant unless it really is some huge sum. But I don't see how we will ever get to that point. And Atlas—*

Harold   *Yeah, I don't see how unit cost can be that great. We're not going to be able to fool around much inside Atlas at this late date.*

Marco   *We ought to think about what we can do without going inside.*

Hans   *On the other hand, if we do convince them that they have to move the paper feed, say, it is going to get costly.*

Harold   *In terms of engineering change but not in terms of unit costs. We still aren't going to go in there with some exotic machinery. All those options, except maybe the E&M device, are just bending metal, cams, gears . . . mechanical stuff, nothing fancy.*

George   *We might have a problem holding tolerances. Machining can get expensive. We ask too much of my people, even with the mechanical parts.*

Sergio   *Maybe we make that another category, another criterion: "Engineering change," "Extent of engineering change."*

Harold   *What you really want to say is something like "Compatible with existing product." Like the QWP we know will work fine. It does in Mars, but we know it will be extremely hard to fit it in Atlas, so . . . Or the E&M, that's going to require a power supply, right?*

Fritz   *But the QWP is our reference. That's not a good example. And, for that matter, what good is the criterion if we know the QWP won't fit? If that's the case, won't all the options be scored a plus, all the same?*

Sergio   *Good point, good point. But I see some that will be just as hard to retrofit—for example, the cam with a solenoid. Solenoids aren't any miniature electronic device. They've got to have room, especially with the forces and reaction times we're going to be demanding.*

Hans   *And the air knife requires a plenum, or the E&Ms—Marco, was it you who said they will need a power supply?*

Sergio   *Fritz, you have a good point. but let's put it up there for now. There won't be maybe any negatives there, but still . . . OK? How did you say it?*

Harold   *"Compatible with existing product" or maybe we ought to say "products," with Leonard in mind.*

Sergio   *Yeah, got it.*

Fritz   *That brings up another thing. Who are we making this design for? Leonard out in Colorado and Atlas are not in sync. Atlas is well along, they're getting into the panic mode now. But Leonard has more time, another year at least, right?*

Sergio   *I spoke to Leonard yesterday, and even though he has another year past Atlas, he wants to see a solution to what he thinks is his dropout problem well before that. He doesn't want to go the panic route.*

Fritz   *But we still have more time with him. And shouldn't we be thinking about the long term?*

Sergio   *We can't afford to do too much of that. I've got the higher-ups breathing down my neck to get something going here. That makes me think of another criterion: How well can we meet a schedule? Let's say "Ease of schedule."*

George   *How about "Pain and suffering"?* [Laughter]

Sergio   *No, we want to be positive about this.*

Marco   *Yeah, so we can mark them down.* [Laughter]

Fritz   *That's why he chose the QWP as a baseline. He knows that can't possibly fit here.*

Sergio   *Come on, guys. That's not true. Let's get serious. We want to get out of here by lunchtime. Jeez, is it already 10:30?*

Hans   *I've got 10:40.*

Sergio   OK. So far we've got—

Harold   *I think we're missing a big one. You all know how difficult it is to keep the QWP clean. Anything mechanical you add in there is going to collect sludge. Some of those, like the cam, are going to have a real problem there with that—keeping clean.*

Sergio   *Good. That's another good one. The guys in Service are not going to like it if they get called out every week.*

Marco   *Does that figure into the cost, the cost of servicing? Do we need a separate category?*

Sergio   *I think we ought to break that one out, just like we did with the paper. That's something we are liable not to think of—what it takes to maintain the fix in the field. So let's add—*

George   *That's going to correlate strongly with "Compatible with existing hardware," I bet.*

Hans   *That may be, but not for all cases. For example, I can see the E&M device fitting in, sliding in nicely right under the bed, hardly disturbing a thing. Yet we don't know what will happen in the field, how often it will have to be serviced.*

Fritz   *We don't even know if it will work.*

Sergio   *We got some interesting results yesterday with a mock-up. I think it looks promising.*

Fritz   *But still, it's got a long way to go. That's what I mean. We don't really know if it will work, and I, at least, can't make a good judgment even though you may be able to, because I don't think we understand enough about the problem!*

Marco   *I'm with Fritz on that. I don't think we have enough information about these different options. I'm finding it hard to do this method, and I think the reason is because we don't really understand the problem.*

Sergio   *How much do we need to know? I admit that the E&M is a long shot, that we've got to get it going, that it will take a longer time to evaluate than, say, the cam concepts, and we've been promised a machine for next week. When we get the hardware, we can do both, evaluate the E&Ms and, in the process, get a firmer grip on what is the problem. But we don't have all year. Jeez, it's 11:00. We don't have all morning either. And besides, this is just an exercise; we are not going to pick a definite option and go with that. We only want to narrow the field some this morning. Then we give it a hard look again, after we've done some work on the three, come back at it and evaluate again. In fact, I can see us running pretty far with, say, two or three options in parallel, as long as they don't interfere. Maybe that's another thing to consider.*

Hans   *Seeing what time it is, maybe we better cut off our criteria here. Serge, I think we better get to ranking.*

Sergio   *OK, OK. So far we've got "Does the job," "Sensitive to paper," "Cost," "Compatible with existing hardware," "Ease of schedule," "Ease of maintenance." Anyone think of any more?*

George   *How about "Ease of production"?*

Marco   *That's in cost. I see that as a main factor in cost.*

Fritz   *Look, I think we have a problem with these criteria. I'm having a hell of a time keeping them straight, trying to fix what they might mean. Are they all to be considered as having the same priority? I still think this exercise is not useful unless we know more about what we have to do, what the problem is.*

Marco   *I think even then these criteria would get all mixed up. When we say "Do the job" I see costs, sludge all in that, too.*

Sergio   *We are always going to have that problem. Where we are now, we've got to move. All I want is to get us narrowed down.*

Fritz   *But you yourself think PT's additional option is worth keeping. I don't think we're ready.*

Sergio   *It's getting late. We're not going to get there today, that's clear. I'll tell you what. Can we meet again?* [Grunts, groans]

Sergio   *No, I promise you. In the meantime, Hans and I will go back and sort out these criteria, try to explain what we see as what they are meant to measure. At least in that way we will start on the same wavelength. I will send you that before we get together. Then we will narrow.*

Marco   *When? I've got to go out to Colorado next week for two days. Can you take that into account?*

Fritz   *And I'm tied up in the lab the early part of the week.*

George   *We've got a production trial scheduled sometime.*

Sergio   *Look, I'll have Cheryl survey, but it might have to go another week. I've got to get out and back to Colorado myself sometime next week. OK? Is that it? That's enough!*

This meeting was later referred to by Sergio as "the disaster meeting." To him it was a failure; he had wanted to establish a set of criteria and then evaluate the options, which he had collected over the previous three months from many of the very same people in the room, against those criteria. His goal was to narrow down the number of options to two or three of the most promising. The Pugh Method seemed a useful tool.

Yet the method didn't work. Why did it not go as Pugh would have it?

One possible reason might be the lack of a clear statement of performance specifications at the start of the meeting. The method assumes that these are given and agreed to by all participants. The fabrication of a list of criteria by which participants are to evaluate the options would then appear to be straightforward. Indeed, the performance specifications themselves might be taken as the evaluation criteria. Evidently, the specification Sergio and Hans had in mind—that the design must solve the dropout problem—was not sufficiently detailed, and as a result the group thrashed about all morning on the issue of meaningful criteria.

As sound as this explanation appears, I am not sure that it is valid. I am not convinced that a more comprehensive and detailed statement of performance specifications at the outset would have enabled Sergio to accomplish his goal. The image conveyed by Pugh is that performance specifications are sacrosanct, in that they are "given," imposed, or not open to question, and, just as important, they are "clear." This is at odds with what I have observed. While performance specifications may be given at the outset of the design process, they are never clear in the sense that all participants in the design process will read them in the same way and, consequently, agree on their relative importance. Any one specification will prompt different readings on the part of different participants. Although the Pugh Method allows for a multiplicity of characteristics in its listing of criteria, the method does not allow for different views on the interpretation of those criteria. Rather, it presumes a single, coherent reading.

This is just not possible, especially early on in the design process. If there is agreement on performance specifications, it is probably the case that the participants have agreed to "apple pie and motherhood"—some self-evident, easily accomplished features—and not to any specifications of substantive content, that is, those that call for serious negotiation. Dropping the passive voice of object-world discourse, we ask, To whom is the statement clear? By whom is it made?

My assertion, then, is that even if Sergio had laid out a set of clearly worded performance specifications at the start of the meeting, the morning would have yet been taken up with an energetic, and possibly even hostile, discussion of their implications, meaning, and relative importance.[2]

The same ambiguity envelops the fourteen options. They are not "out there," uniquely defined and understood by each and every participant in the same way, just waiting to be evaluated against some clear performance specifications. What is out there? There is a list of names, one label for each option, and some sketches indicating the essential principles of operation of each in turn—iconic diagrams of underlying form as much as sketchy representations of a physical artifact. There exists hardware from past and current projects. Some participants have a more detailed understanding of these precedents than do others. Some think of the way in which the competition uses one or another method for preventing the degradation of an image in the printing process.

All use special words and phrases in their proposing and questioning—dropout itself, E&M device, paper path, pretreatment, tolerances, sludge, power supply, QWP—but here, too, one ought not assume that they mean the same thing to each and every participant.

Yet while the method did not work as Pugh would have it, a useful exchange did take place. In naming options and their features—some desirable, others not—and attempting to articulate a set of independent criteria, the team wrestled with what it projected as concrete options for the design of a fix to the dropout problem. In this process of getting the names right and developing shared meanings, they were constructing the objects of design as much as classifying a collection of clearly defined artifacts with reference to a set of given, clearly understood criteria. Indeed, they were already designing, making choices and tentative agreements, laying the groundwork for what, at future meetings and within object worlds, would be considered legitimate, acceptable, and determined.

The use of a two-dimensional matrix, in an attempt to manage and control design complexity through its division and dissection into what can be made to appear as independent options and features, is common practice in engineering. It is an attempt, not without merit, at analyzing objects that show no hierarchy or preferred relationships to one another.

One might think it would be adequate to make a list of design options and then append to each entry a list of the criteria it satisfies. But this wouldn't do. The matrix explicitly shows the options and criteria in relation to one another without bias, on a level playing field. Comparison can then be made, row against row or column against column, freely in all directions.

This opens the door to some confusion, as shown at the meeting. The setting of criteria for classification is intimately wrapped up with the problem of classifying itself. In effect, the group is using the options, to which they wish to bring order, as the source of criteria against which they will classify these very same options. No wonder they appear to go around in circles. We might say that the options produce the criteria as much as the criteria define the worth of an option. The cells can be taken as the locus of the intersection of object as design concept and object as criteria.[3]

In the light of all this, the meeting was not a disaster. Designing was happening. From uncertain beginnings such as these, participants in design add to, manipulate, and transform a customer's expectations and, in time, develop a shared vision of the artifact—how it is to be made, how it will perform, how much it will cost, even how to fix it if something goes wrong. "Shared vision" is the key phrase: The design is the shared vision, and the shared vision is the design—a (temporary) synthesis of the different participants' work within object worlds. Some of this shared vision is made explicit in documents, texts, and artifacts— in formal assembly and detail drawings, operation and service manuals, contractual disclaimers, production schedules, marketing copy, test plans, parts lists, procurement orders, mock-ups, and prototypes. But in the process of designing, the shared vision is less artifactual; each participant in the process has a personal collection of sketches, flow-charts, cost estimates, spreadsheets, models, and above all stories—stories to tell about their particular vision of the object. The shared vision, as some synthetic representation of the artifact as a whole, is not in documents or written plans. To the extent that it exists as a whole, it is a social construction—dynamic, plastic, given nuance and new meaning at each informal gathering of two and three in a hallway or at formal meetings such as scheduled design reviews.

The thesis of this book is that the process of designing is a process of achieving consensus among participants with different "interests" in the design, and that those different interests are not reconcilable in object-world terms. There is no overriding perspective, method, science, or technique that can control or manage the design process in object-world terms. The process is necessarily social and requires the participants to negotiate their differences and construct meaning through direct, and preferably face-to-face, exchange.

Designing and design decisions depend, then, upon the values and interests of participants. This is not to deny the importance of scientific and technical constraints and specifications, but these are not determinate. Participants must move beyond the secure confines of object worlds and engage one another on more common and less ordered ground for design to proceed. In this, participants' interests shape their proposals, explanations, and understandings.

The word *interests* needs definition because it can easily be misconstrued. Reviewing the discourse of the disaster meeting reveals the sort of conflicting interests to which I am referring. The participants each have their own way of looking at "the problem," and they are each interested in having their concerns given due consideration. These concerns derive from their technical expertise, experience, and responsibilities. This is what I mean by the interests of participants: their concerns, rooted in their knowledge and belief about the nature of good design practice within their respective object worlds, about what constitutes the design task, or indeed, what constitutes the problem. The concerns, visions, and interests of different participants are often in conflict, as is evident if we return to the so-called disaster meeting.

Fritz speaks of "not knowing enough about the problem." He is a chemist who works in the Research Division of the firm. His interest is in understanding the problem thoroughly. He knows that in good basic research practice it is important to expend time and energy to ferret out the intricacies of every question. Why is dropout occurring? He wants to come to grips with that question, to test and systematically probe the system and diagnose this degradation in performance in terms of the basic underlying chemical and mechanical phenomena and mechanisms. This is what he sees as the problem whose solution is prerequisite to design. The "failure" of the Pugh Method is no surprise to Fritz. His mistrust reflects his interest in doing the job right. For him, that lies in constructing a convincing and complete definition of the phenomenon, not in applying some managerial snake oil or turning the crank of a machine with the expectation that it will produce a best option.

Sergio sees the problem as meeting the needs of two product designs that are in trouble and whose success in the marketplace is now in jeopardy. His interest is in designing a quick yet robust fix. He doesn't have time to try to understand every detail of the problem. He values time. Understanding is sufficient if it provides a framework for the rational redesign that will meet Atlas's needs. The schedule is very tight. He doesn't have the resources to study the problem to the extent that Fritz would like. Besides, he claims you can overdo it and never get anywhere if you insist on complete confidence in your design before you start "cutting metal."

Sergio often complains about the expectations placed on the Advanced Technology Division, expectations that go beyond their stated mission. Why is it that programs come to them at the last minute for help? Why is it so difficult to obtain the resources needed to support the tasks they are asked to undertake? He confides in me:

The organization stinks. I'm tired of this runaround. Management is the problem; they don't know what they are asking us to do.

Sergio sees the Pugh methodology as a useful, rational method for resolving a problem that has a complex managerial as well as an engineering component. Certainly it is not a scientific problem.

Marco, like Fritz, is a chemist, but he works on the development of new products, including at the moment the successor to Atlas. He sees chemical pretreatment as the cleanest solution. The mechanical devices that he hears discussed at these meetings seem like so much medieval machinery.

George, from Production, raises his voice when he sees that what's on the table is going to be difficult to fabricate; to "hold tolerances" to the specifications some are suggesting will make the machining of parts very expensive.

Harold like George is from Atlas. He is concerned that whatever the group proposes not require any great change in the current configuration of "his" hardware. He uses other words to talk about the problem. He speaks about "particles" where Fritz or Sergio would say "dropout"; *particles* is an older word, from last year's dropout problem. His colleagues in Product Development over at Atlas now speak of particle-centered voids (PCVs) as the problem they are trying to eliminate. Now *particles* originally referred to the microscopic particles that sometimes found their way into the chemical solutions used in processing the print. Harold, then, saw the problem as eliminating these microparticles, filtering them out of the system. He, like Marco, favored chemical pretreatment.

To PT the problem was due to the paper not being held snugly against the bed during processing. He claimed if the paper were so held, it didn't matter if there were microparticles in the chemicals; they would not be able to engender the visible "hole" around them. To him the challenge was to design a way to hold the paper flat against the bed. He had done an analysis in support of a prior product development effort, but his work had gone unused.

In this case the differing interests of participants in design have been revealed through discourse at a concept selection meeting. They also surface in informal conversations among two or three people in the hallway and in one-on-one interviews. They appear, too, in the artifacts participants produce in design—in the way a prototype is fashioned, in the way a test is conducted, in the way details of a drawing are attended to, and in the way cost estimates and schedules are prepared.

To see interests, values, and norms in these latter activities requires a framework that allows for variability as part of design work. It also presumes that this variability is recognized by, and meaningful to, members of the subculture and that at some level the interests of individuals reflect shared norms about what constitutes a good design. These, like performance specifications, are not given from above, learned in an ethics course at school, inherited, or genetic in any sense of the word, but are mutually constructed and maintained by members of the firm. Norms and interests in themselves are social constructions of participants in design.[4]

Let us return to Sergio's meeting. Lurking in the background of the different perspectives and interests displayed by those in attendance is the problem of defining *dropout* in quantitative and instrumental terms. How is one to measure dropout? How do you make a standard measure of a blemish in a photographic print when the blemish ranges in size from microscopic to millimeters and when perception of the blemish is very much dependent upon the ambient conditions surrounding the print viewer?

One can set standards—perhaps agreeing on print images that will be used for testing under some standard lighting conditions and then count and construct a frequency distribution of blemishes within different ranges of diameter—but these standards then become part of the problem. What should the standard be? This is problematic since it must include an agreement on what participants involved in setting the standard think the user or the customer will say is a blemish in a print. Indeed, at subsequent meetings, different participants presented different test prints for discussion. Their meanings had to be negotiated. They had to agree to a specification: no more than a number x of dropouts with diameters within a certain range. And they had to agree that no more than y prints out of a run of 1,000 could exceed this specification. All of this setting of specifications requires negotiation of different understandings of the problem, of ways of saving costs, and of getting the job done on time.

A critical question in this process of standard setting is how to tell when the problem has been "solved." Attempts to answer this question might very well lead to a claim that nothing need be done! That is, in practice, the quality of most of the prints produced by the current design, the machine as is, is acceptable; the customer does not see a blemish or so rarely identifies a place on the print as a blemish that no blame is laid on the product or its makers. This is not to say that the customer might not see the print produced by a competitor's print machine as better. He or she may note something lacking in Atlas's print relative to the competition's, but doesn't articulate this as being due to a microscopic blemish. The customer does no standard testing but simply expresses a "feeling." Now one might suppose that dropout is the source of that feeling. But then again, it might not be. There might be other ways in which the competition's product is better.[5]

To what extent does Sergio's group discussion of dropout presume a unique, shared understanding of the problem—some generally agreed-upon malaise involving Atlas's potential to produce a quality print? If it is not unique, then other formulations of a (*the*) problem, and significantly different formulations in terms of their implication for design, exist—including the possibility that nothing need be changed at all. At this point we can see that the definition of the problem is a social choice. Furthermore, it logically follows that if we admit the possibility of problem varieties—a garden of varieties of the species *dropout* (it is interesting to note that as standards were developed, different participants used different test prints)—then we can legitimately ask if there is "really" a problem at all.

The question may be put thus: To what extent is dropout a social construction? The view here is that because participants hold different views, espouse different interests, and work within different object worlds, the reconciliation of these differences in the definition of the problem is not unique in any instrumental, operational, or scientific way. Dropout, as a problem, is thus a social construction.

I will add a further provocation: The possibility that the problem of dropout could be defined otherwise (or as a nonproblem) remains even when the team has agreed that the problem is solved, that is, after the fix has been made. Participants in design might agree that the malaise has been eliminated (if *we agree* to see according to this test pattern, if *we agree* to use this range of paperweights, if *we agree* that the costs are appropriate, and so on), but there is no way to establish the solution's validity for all possible conditions of testing or use.[6]

This is not to say that what participants see, define, fabricate, and do on their way to a solution is irrelevant or that one problem definition is as good as any other. For, while the uncertainty and ambiguity that prevail in design allow the sort of indeterminacy advocated here, there are constraints, of tradition as much as of science, on the visions, conjectures, and refutations of participants. What matters is that participants gain and remain in control of what they construe as the problem, working both across and within their respective object worlds.

If we take the perspective that designing is a process of negotiation and exchange across different interests, object worlds, and disciplines and that participants must work to establish and maintain both the problem and the norms to be engaged in judging their contributions to the design task, then we can see Sergio's meeting not as a failure but as a first engagement on the road to the design of a fix of the (of *a*) dropout problem—albeit a rough and tense first step.

Sergio's objective was ostensibly to define criteria that the team could use to eliminate all but two or three options. While they never got to the evaluation part, team members were continually referring to specific options in describing what they meant by a criterion. We hear them exploring and evaluating how much power the E&M device would require, how the air knife would require a plenum (later a decisive factor in its elimination). In the course of discussing specific options, they engage in a spirited debate about the nature of the problem, render some underlying form concrete, set boundaries on some of the options' possibilities, and work over what they perceived were important constraints, organizational as well as technical. In this way the meeting prefigured the object of design, and the team members were designing.

Still, it is a fact that Sergio did not get agreement on a set of performance specifications or criteria for judging the options. Indeed, at times it appeared that his project would fall apart—the calls to go back to the lab sound particularly discouraging. (And do what with Atlas's problem? Something has to be done. Or does it?) Yet, though the meeting failed in Sergio's and in object-world terms, participants went off with a better sense, shared to some degree, of the strong and weak points of the fourteen different options. There was no documented agreement that two or three ought to be pursued further and the remainder rejected. That was yet to come.

And it did come. Sergio did what he said he would do. At a subsequent meeting he laid out a set of criteria, essentially the set developed at the disaster meeting but reshaped to reflect the interests of the more vocal participants, clothed in more rational and instrumental garb. He was able to get agreement from those at the second meeting with relative ease. By that time participants had better defined their own roles vis-à-vis the dropout problem. The meeting was less contentious due in part to the fact that almost no one at the meeting, except Hans, was going to be working directly on the project. So things were calmer.

Sergio was also better prepared. He had a better sense of what it would take to get consensus. While there were still some complaints that "they ought to really understand the problem," he convinced everyone that the problem had become critical in the intervening two weeks. He had learned from his trip to Colorado that Leonard was now fretting over the possibility of dropout, beyond what he had originally expressed. So Sergio could claim that they must do something in short order.

At a subsequent meeting Sergio obtained consensus to go ahead with the E&M option, the chemistry of pretreatment, and the paper-path change. The air knife had been eliminated by Don in the intervening week.

Module Voltage

I want to turn to a second example of design discourse, this time drawn from the design of the residential photovoltaic module. We intervene at a point well along in the process when the physical size of the module had been "frozen" and crude values for the number of cells and the module's power rating had been set. Still to be defined were the frame structure, the networking of the cells, the number and placement of junction boxes, the number and location of protective diodes, a production schedule (though a production goal had been set), and marketing targets.

Members of the Panel Fabrication group and the Marketing and Systems Engineering group are gathered to set the design of the module junction box. At least that is the one stated agenda item. Those assembled for the meeting include William, the head of Cell Production; Brad, the head of Marketing and Systems Engineering; Steve, the chief of Panel Fabrication; Ed and Tom, engineers from the Systems group; and Jane, an engineer from Quality Control who is responsible for all cell and module testing.

At the outset their conversation ranges widely and includes what, at first hearing, sound like features having little to do with the design of the junction box.

Jane　*What cell size are we going to go with? Have we fixed on that yet?*

William　*Brad, what have we promised our customer in the way of module performance?*

Brad　*We are verbally committed to a 70-volt module, but there is some flexibility there; we're not set in concrete.*

William　*When did we promise delivery?*

Steve　*We can't do much until we have the glass on hand. When will the superstrate be shipped to us?*

William　*How did we set the overall size . . . external dimensions?*

Steve　*Let's get back to business. Where will the diodes be located? We need an engineer's layout.*

William　*Can we integrate them into the module? . . . a neat package.*

Ed　*Yeah, but how many? JPL* [Jet Propulsion Laboratory] *sets a spec at one every twelve cells.*

Steve　*Yes, but what panel voltage are we talking about?*

Brad   *If we went to two junction boxes—*

Ed   *Look, we've got to know a voltage. Tom, where is that layout?*

Brad   *A 36-volt module would be ideal for one every twelve cells.*

Tom   *Yeah, but where do we locate the junction box. How do we tie it down?*

Steve   *Are we going to frame the thing? You know if we build a 12-volt panel, we won't need diodes at all.*

Jane   *How many cells in series to get the twelve volts?*

This sequence of fits and starts, of probes and parries, ought to be read as a reorientation of participants toward the object of design, the residential module as a whole. Some are statements of fact, some are questions, some are mixed. Some probes are true queries in search of information; others are more rhetorical and critical; and still others advance a design change or suggest a new feature. Participants are thinking more broadly than about the design of the junction box during this first stage of their meeting; they are reaffirming their visions and understandings of the design of the object as it stands at the moment.

Participants in this exchange display their varied responsibilities and perspectives. To those in Panel Fabrication, the *panel,* not the *module,* is a collection of fragile silicon cells, strung together by what appear as thin leads of tinsel, sandwiched between a glass superstrate and a backing of tedlar sheets and metal foil, all fused together in a controlled heating process. They worry about blistering and weathering of this protective backing and about the subsequent corrosion of the tinsel, if that should occur. They worry about meeting production goals and keeping the fabrication process running and under control. They wonder about whether a frame is needed at all.

When people from Cell Testing look at the firm's product, they see the assemblage of fragile silicon wafers not as an arbitrary collection but as a set of cells carefully cultured and grown, doped and cut, tested, sorted, and finally grouped together in batches and sent off to be strung together in a panel. Their concern is with the careful treatment accorded these slender blanks of crystalline silicon and the fine control that must be maintained in the growing process if the cells are to have consistent quality.[7]

To those in Systems Engineering, the photovoltaic module is itself a closed unit. (Note the use of the word *panel* in Panel Fabrication, and *module* in Systems Engineering.) While the fabrication people see their panel as a physical flat plane, much like a piece of glass—a thing worthy of complete attention in itself—the systems engineers see and speak of the module as a functioning, power-producing element, just one of many building blocks to be assembled into an integrated system of considerable power. They worry about the "balance of systems," about what batteries will be required for energy storage, about what control strategies will get the most out of the system for a given module efficiency, and about the costs of the whole. Even the blue of the cells is seen differently: To those in Cell Testing, variations in shades of blue indicate variations in cell performance. Systems engineers read the blueness as an aesthetic quality of the field array as a whole. The same artifact; different objects.

About the only commonly shared characteristic of the module before the meeting began was that it was to be four by six feet, to match standard construction practice, and the rated power was to be in the vicinity of 200 watts. Although an individual cell size had been set, there was a certain amount of flexibility in this. All expected that the output of the module would feed into a power conditioning unit, an "inverter," where it would be transformed from direct to alternating current. All else was subject to change, definition, and negotiation.

Before moving ahead and deciding what they were assembled to decide—ostensibly the features of the junction box—participants construct scenarios of how some ingredients will function relative to others. In the telling of these short vignettes, they draw on their individual object-world knowledge, articulating a justification for their proposal or critique of someone else's in a way they expect will convey meaning to all at the meeting. But they can never be sure of a sound hearing; in this first phase of the meeting, we hear people frequently rephrase and restate their concerns.

In time, one question emerges and holds their collective attention: What should the module voltage be?

In retrospect I can reconstruct a trajectory that shows why module voltage became the preeminent focus of further discussion. But it could have gone otherwise. That is, module voltage might not have emerged as a priority. The number of diodes and their location in the electrical network, the number of junction boxes and their positioning, or even the necessity or not of having a supporting frame—any one of these features might have served to fix the course of the discussion and move the design process forward. In object-world terms, all of these features are interrelated: The design of the junction box (or boxes) is related to the number of diodes, the module voltage, the number of cells in a series string, the physical arrangement of the cells within the panel, the location of the junction box(es), the structure of the frame, and so on. If one or another of these features was chosen as a focus instead of module voltage, the latter would in short order become an ingredient of the deliberations along with the others.

For example, the diode's function is to control the direction of current flow in the module. Each diode is about the size of a quarter and functions like a gate, allowing the current to flow in only one direction. Without the diodes, the cells could overheat if there were a failure in an electrical connection at any one of the hundreds of connections among cells. Of course, if there were no failures, the diodes would not be necessary. Moreover, since they themselves dissipate power and represent an inefficiency, the fewer of them there the better. How many to include in the network is an open question whose answer depends on designers' reasoning about the likelihood of a failed connection and the way the cells are connected—the topology of the series and parallel connections.

Now this topology determines the nominal module voltage. Furthermore, since the diodes are best placed in a protective enclosure (although the possibility of integrating them into the panel was suggested), the location and number of diodes will enter into decisions about the location of the junction box (and the number of boxes). And because the junction box requires a support, we can claim that the design of the module frame structure also depends in part upon decisions made about diodes. Everything hangs together, including the choice of materials and fabrication process for the module substrate: Without watertight security, it is likely that weathering will eventually cause a failed connection.[8]

Whether starting with diodes or some other feature matters—that is, whether the final design would be different if a different trajectory were followed—is difficult to argue. However, our inability to respond to this sort of counterhistorical question does not negate the possibility of alternative paths. There is always a good bit of choice available amid the uncertainty of designing. At a design review meeting or, indeed, within the confines of work within object worlds, there is ample room for maneuvering images, recasting assumptions, and enhancing labels to extend their reach and significance. Module voltage serves as a vehicle, a "cover term" that helps structure the discourse.

We return to the meeting:

Steve *If we go with 12 volts, we are in good shape with diodes. We don't need to have any at all.*

Tom   *I don't know. I don't like the idea of none at all.*

William   *But 12 volts is awfully low. Brad, you say you promised 70 volts.*

Steve   *You want 70 volts, then you string five panels together. I take it that 70 is an open-circuit value.*

Steve's interest is in a low-voltage module. A high-voltage module would require "bypass" diodes, as stated in a recommended specification from the government-funded Jet Propulsion Laboratory. They agree with the spec: A high-voltage module ought to have bypass diodes. But it is unclear to them how to network the cells simply and with as many diodes as the JPL recommends to get the high-voltage that Systems Engineering wants.

But why is a high voltage required? A low, 12-volt module could be built without any diodes whatsoever, by networking the cells in parallel in a unique arrangement. Fabrication in this case would be straightforward; simplicity, an often-touted design norm, would prevail. Location of the junction box would also be simpler, and the selection and matching of cells would be less of a chore.

Steve attempts to extend his vision to encompass the interests of other worlds within the firm: With a 12-volt design there is more cell paralleling and, consequently, tolerances on cell production need not be so tight; this means lower production costs. With a low-voltage module as a fundamental building block, a wide, almost continuous, range of *system* voltages (such as Brad's 70 volts) could be obtained in the field. In this way Steve seeks consensus centered on his own preferred design.

Ed   *That 70 is not low. Look, were going to have to meet the inverter window. Tom, what's that . . . around 260 or something?*

Tom   *In that neighborhood. I got the specs on American Power's unit right here somewhere.*

Ed   *Now we are talking of stringing something like eighteen modules at 12 volts. That means at least a $3^1/_2$ kilowatt array at a minimum.*

Steve   *Sounds about right to me for a residential power system. What do you think, Brad? What do your marketing studies say?*

Brad   *I think that's high.*

Tom   *You mean 70 volts?*

Brad   *No, I mean setting a minimum rated power for a residential system at $3^1/_2$ kilowatts.*

Steve   *That looks to me like it would be just right. Produce about 450 kilowatt-hours on my rooftop each month on the average—*

Ed   *Yeah, but what if you want 20 percent more power. You've got to add another string of eighteen modules.*

Brad   *That's clearly a problem. Twelve volts is not going to work.*

Steve   *But with a 70-volt panel. How are we going to network that? That looks like an odd configuration.*

Tom   *It's not too bad. I can see a couple of ways to go.*

Steve   *Including diodes? That's the kicker. I can see a network, but we're talking five, no ten, diodes.*

Jane   *Brad, is 70 that tight? Why not 24?*

Ed   *The diodes aren't that bad. In fact, if we go to 96 volts, the picture improves. There are more options.*

The systems engineers make a case for a high-voltage module. This would allow them to match the input requirements of the power conditioning subsystem with fewer modules strung together in series. They note the lack of flexibility in the internal wiring of a 12-volt module, even though they have to acknowledge that the lower-voltage module would give them more flexibility in varying the overall array voltage.

Brad, the head of Marketing and Systems Engineering, also prefers a high-voltage module, claiming advantages in meeting potential customers' needs. He has economic studies to back his point. Furthermore, another government laboratory with responsibility for promoting the development of low-cost photovoltaic technology wants to test a high-voltage, large module. Here is another customer it is best not to ignore.

Brad   *I think we ought to try 48 volts. That's a good voltage for us old radio aircraft people.*

At this point, Ed goes up to the board and sketches out a possible network that would give a module voltage of 48 volts. It appears as a simple transformation of the 96-volt networks that he and Tom had been playing with all week. In the process they had convinced themselves that they could get away with just eight diodes. Tom joins Ed at the board, and together they work in the diodes. Junction box placement seems to follow effortlessly.

Steve, from Panel Fabrication, agrees that the design looks reasonable. The junction box is located in a good spot; it might even be possible to use two smaller boxes symmetrically located in the center of the back of the panel. Steve likes that option.

In five minutes the whole pattern of the design has come into focus. Here is a detailed topology that fixes the module voltage and opens pathways to the definition of all the other, related features. This is the hard stuff, and the right stuff, all there in object-world terms, complete with a prescribed number of diodes—a number consistent with JPL's recommendation—and with the junction box locations specified well enough that Panel Fabrication can see how they can fix them securely to the substrate. It also includes a suggested glass size; and framing of the module appears to be possible without the complications of supporting the junction box. The team has constructed a common vision. There is banter and laughter, then closure and scheduling of another meeting to firm up the details.

Yet, after the meeting, questioning key participants, I discover differing interpretations of the significance of the choice of 48 volts as a module voltage and differing accounts of how they came to make this choice. In part, these differences derive from the variety of meanings that can be ascribed to the term *module voltage*. Module voltage can be *open-circuit voltage*—the voltage measured by a voltmeter when the module sits alone, disconnected from any circuitry, that is, the voltage measured across the module's *open* terminals. Module voltage can be *max-power voltage*—the voltage at which the power produced by a current flowing through the module and out around through a closed circuit is a maximum. Module voltage can be *operating voltage*—the voltage one might expect to see with the module working in parallel with a battery to supply power to a load. And module voltage can be *nominal voltage*—the voltage defined by the number of cells in a series string, attributing $1/2$ volt per cell in the string. A nominal 12-volt module could include anywhere from thirty to forty cells in a series string.

All but the last of these definitions are context dependent in the sense that they require you to envision the module as part of an electrical circuit (or not, in the case of the open-circuit voltage) and in operation before taking any measurement.

In all cases, the actual measured value will be a function of the temperature of the module as well as the intensity of solar radiation. The $1/2$ volt per cell open-circuit voltage is the value one would measure with the temperature around 20°C on a clear day, at noon, when the intensity of solar radiation impinging upon the flat surface of a cell that is perpendicular to the sun's rays would be approximately 1 kilowatt per square meter. Cell temperature and solar intensity must be specified if the numerical values measured for open-circuit, max-power, or operating voltage are to be understood correctly. To ensure consistency, *standard conditions* of cell temperature and solar intensity are often prescribed, but at the time of the junction box meeting standard conditions were still in a state of flux. Discussion was ongoing about what temperature would best serve as a reference: Should the temperature of the module or the ambient temperature of the air set the condition? (The module runs hot when illuminated by the sun.)

At the meeting, module voltage was defined crudely as nominal voltage in the systems engineers' network drawing. Other readings were in the air, however, as indicated in the exchanges throughout the meeting, but these were never tied down or made the focus of discussion. It is the openness of the definition that gave participants leeway to reconstruct how they had arrived at the choice of 48 volts and what the implications of this choice were.

Panel Fabrication, for example, saw it as a compromise:

Systems originally wanted a 96-volt module. I would have liked to have seen a 12-volt module. I can live with 48 volts.

They read the 48 volts as a modest excursion from their original 12-volt vision, and the eight diodes seemed reasonable. (Subsequently the number of diodes was increased to twelve, but by that time the module voltage had been frozen and no longer negotiable.)

Systems Engineering, on the other hand, saw the process as one in which instrumental (right) reasoning prevailed over the interests of other parties to the design who did not have a full understanding of market conditions and the technical requirements of integrating the module into a fully functioning system in the field. They reported that what was chosen was a high-voltage module:

It was really a 60-volt module at JPL standard testing conditions.

The term *module voltage*, as nominal voltage, is a label for an arrangement of cells in a network—the network drawn on the board at the meeting. The network is hard and objective. Yet, although everyone "saw" the same drawing and agreed that it was a good representation, participants interpreted its implications and meaning differently. The agreement appears solid only if we restrict our attention to the physical module and draw a boundary around the thing in itself. If we open up our field of view and think about module voltage with boundaries drawn further afield—for example, around the cell production and panel fabrication process, or around the design of a desalination plant in the field—then what that particular topology means (in this case, what module voltage means) will not be so commonly understood. Away from the meeting, during my one-on-one interviews, the boundaries established at the meeting—whose setting was so crucial to the success of the meeting—no longer restrained participants' accounts.

For Panel Fabrication, module voltage in context means consideration of cell matching, junction box placement, numbers of diodes, and the possibility of a failed connection. For Marketing and Systems Engineering, module voltage in context means consideration of the full system, whether it is a stand-alone system, with batteries for energy storage, or whether it is part of a utility grid system, feeding its output into an inverter and operating continuously at maximum power, whether it is to work in a cold climate or in a desert, how to network the modules to get an appropriate system voltage, and even the appearance of the array in the field. Module voltage—one phrase, same artifact, but different objects, contexts, and accounts of what went on at the meeting.

Module voltage functions much like what an anthropologist would call a "key symbol."[9] Participants capitalize on the variety of its meanings. Module voltage names. Once uttered, it calls to mind a constellation of concepts and relationships within various contexts in use and in its making. The umbrella-like, open character of the word provides design participants with the room they need to negotiate differing, even divergent, views of the meaning of a network of cells. Once said, and you have a way into design action. It opens the door, so to speak, to designing photovoltaic modules. Knowing how to use the phrase in a meeting is critical to design.

The words *force* or *energy* show this ambiguous richness in the historical development of classical mechanics. *Module voltage* and other terms like *efficiency* or *reliability* offer the same opportunity for creative design of photovoltaic technology. Creativity springs from around the edges of words.

Module voltage grew up with the photovoltaic industry over a period of years. Naming goes on within more transient contexts in the day-to-day work of design. For example, Hector at Photoquik is working on a new linkage for the redesigned baffle of the print processing bed. He needs a name, something less formal than "parallel motion linkage," something akin to a nickname, an iconic phrase expressing function as well as form—"drive post," "feed link," "grab arm"—to signify the object and make it real. To speak of "reifying" when speaking of "objects" may seem at first redundant, but it is no redundancy. A grab arm doesn't exist as artifact; choosing a label for the object brings it together, consolidates meaning, constructs boundaries and turf, and fixes an identity, all before it exists as artifact.[10]

Choice, however, is not quite the right word. (Choice itself is not the right choice!) *Construction* is better—a construction that in itself is a design act, prefiguring real physical conditions and other design acts. The label brings some baggage with it: anthropomorphic leanings perhaps, analogies and metaphorical implications certainly. The label has to be right. Design participants struggle over words. Naming is designing.

What's an Explosive?

We go now to another gathering, a meeting intended to set the test protocol for the prototype unit of Amxray's inspection machine. The meeting's formal agenda stretches out over a full day. It calls for an Introduction starting at 8:30 A.M. and adjournment at 5:15 P.M. The morning's session is to be devoted to the demonstration project, the afternoon's to the production system. The main task of the morning session is to reach agreement on a list of test articles to be placed in the cargo containers and run through the prototype inspection system. The intent is to verify the system's ability to identify foreign articles.

To ensure that the meeting stays on topic, BG's project director circulated a memo—"Initial List of Test Items"—for review a week before. It proposed a sequence of tests of possible contraband cargoes. The test items were to be disguised and mixed in with "normal" or legitimate goods in each container run through the collimated x-ray beam. One test sequence was to focus on weapons: "a handgun mixed in with a collection of household goods (e.g., pots and pans)." Another was to focus on narcotics: "two sculptures, one hollowed and containing a plastic bag of powder." Still another was intended to test the system's ability to detect hidden explosives: "two sewing machines, one of which contains an explosive device." There was also a catchall, "other contraband" category.

One Amxray executive who had prime responsibility for overseeing the design of the detector system and who knew the system from the level of its physics up to its imaging software had noted on his copy of the memo which tests he thought were feasible and which would not work. He sent copies to, Arnie, the project scientist, Michael, the project manager within the firm, and Jim, the head of electronics, with the notation "We've got to put together a strategy."

Al, BG's project director in town for but the day, led off:

Al *We've a lot on the agenda. I take it you have had a chance to go over my memo? Before we start, I want to emphasize the goals of the demonstration project. I've listed them right on page 2. First, we want to ensure that the performance specifications are met, but that's not all. We want to convince our potential customers, after seeing our demonstration, to purchase the production system. OK. On page 3—*

Arnie *Before we start, I have a general observation. I went over this list pretty carefully, and I had troubles with it. It looks to me like you're asking too much of the system. This looks like the test program for a baggage system.* [There are significant differences in resolution between existing baggage inspection technology and the cargo container system.]

Al *I know we are pushing the technology, but from the user's perspective, this is what he wants to detect, so let's go to specifics and try it out. That's what we're here for this morning, anyway. Let's take a look at the first item. Explosives.*

Immediately, there was a plethora of questions from every direction:

What's in a typical cargo container? What would a terrorist attempt to smuggle across international borders?

There were bursts of discussion of recent terrorist activities, the telling of stories relayed by a customs official at the border, and more questions:

How would they try to hide the stuff? How would they pack it? How much stuff would they put in a single cargo container?

Alternatively, the discussion would move to the system in use, in the hands of a customer:

Are we trying to demonstrate the capabilities of the prototype system alone or the production system as well? And what does the customer *think he wants to detect?*

There were almost as many unknowns on this side of the issue. The customer's interests and abilities were uncertain, albeit not to the extent of the terrorists':

What can we detect? Will that require a trained operator? What can our competition detect? How much is all this going to cost?

The discussion swung back and forth from terrorists and their explosives to physicists and their x-ray detection machinery. Then there was the matter of the explosive matter itself:

How do we simulate explosives? We're not planning on putting a half ton of TNT into my test chamber, are we? What is this stuff? I see a "device" with wires . . . like a bomb. I think we can detect the plastic if the density difference is good so that it contrasts with the background.

This exchange about explosives shows again the play of different interests and perspectives. There are scientific questions that are most relevant to those concerned with the "physics of the device"—insider questions, if you like—and there are scientific responses to these questions in terms of density and contrast. Another participant, however, sees the explosive not as a bag of powder or plastic but as a "device with wires . . . seems to be a bomb, rather than explosives." This extends the context to include what an operator can identify as well as what a terrorist might try to conceal—outsider questions, if you like.

The first issue of the meeting, then, is to agree on what the word *explosive* is to signify and thus what is going to be placed in the cargo container that will test the performance of the x-ray inspection system. This fixing of meaning is not just a question of semantics, although that is how the process appears. Behind the play with names and labels, participants are defining the contraband to be tested. That contraband will yield certain data, certain images on the operator's screen. Image quality depends upon the functioning of all the intervening hardware: the x-ray source, collimator, detectors, signal conversion, and system software. But these are all a matter of design, so the words that label, like *explosive,* and what they are taken to mean in terms of test items are critical in participants' thinking about what each subsystem must be designed to do, how it must function. If *explosive* were taken to mean a bag of plastic, that would present one type of challenge to the existing design. (When the meeting was held, the design was well along. The essential characteristics of the system had been fixed in people's minds for several months.) Hence the concern noted on the memo. If *explosive* were construed as a device with wires, the challenge would be different and simpler.

In this naming process, participants need not change the words. If they agree that "ten pounds of explosives" is a device with wires, they need not change this statement, though certainly its meaning will have changed in the eyes of some.

The x-ray design team is grappling with a question very much like the one Sergio's people faced in defining criteria against which to judge their fourteen design options. And there are clear resonances with the residential PV module design team's negotiation of module voltage—the problem of setting clear and concise performance specifications at the outset of design. While the initial statement about "ten pounds of explosives in a cargo container" appears specific, it is not so meaningful, or, rather, its meaning has to be constructed.

In this, different readings are evident. The scientist speaks about the possible resolution and discriminatory powers of the system, the project director worries about what will impress potential customers of the production system, the chief of Test Operations asks where he is going to get all this stuff (real explosives?) to put in his test chamber. Even the terrorists have their interests taken into account in all the second-guessing about how they design explosives and how they might try to conceal them in a cargo container. In their negotiation of meaning, participants set boundaries, fill in and firm up constraints on the design of the whole system. Again, naming is critical to design.

Consensus is essential to this process, but only to an appropriate level of detail. Giving meaning to *explosive* is not a matter of extending its reach outward, linearly, or branching off, adding more and more specifics; that is an object-world rendition of what happened. Rather, participants labor to get the structure right. They must be sure they are working with a common, albeit loosely fitting, skeleton before going off and adding flesh to the bones.

The session comes to a close with the agreement that an explosive is a device with wires. As at Solaray's junction box meeting, names have been identified and embellished in ways that work both for the insiders with their interests and knowledge about the physics of the device and the outsiders with their interest in the system working in a marketable context.

Here again we see how, even in a discussion about hard stuff—technical apparatus, instrumental operations, and inanimate things outside of us—a healthy measure of ambiguity and uncertainty makes room for designing. The language of a performance specification is not precise. The implication these labels have is never fixed for all time or for all contexts. In process, it is always possible to invalidate yesterday's design move because the object the language points to does not exist, or, better yet, it only exists within participants' individual object worlds. Within one participant's thinking a definition may be precise, but in another's world precision may be altogether lacking. Only after the fact, when design yields to artifact, do meanings appear firm and consonant. The reality of the artifact, read in retrospect, can lure you to think otherwise, but that is a *trompe l'oeil*.

In process, the object masks the uncertain and the unknown. Participants envision and construe the uncertain as options, but behind the mask the unknown lies waiting—and that, too, is valued by participants. Uncertainty is what gives life to the design process and makes it the challenge that it is. If the process lacks uncertainty, then you can be sure it is not designing but copying.

The objects of their instrumental discourse do exist as representations within participants' object worlds, and there one can of course speak in instrumental terms. But different object worlds are not congruent, and so the instrumental stories of object worlds are not of one piece. Herein lies the source of ambiguity.

This, too, ought to be valued. Ambiguity is essential to design process, allowing participants the freedom to maneuver independently within object worlds and providing room for the recasting of meaning in the negotiations with others.

Others have expressed similar thoughts. Here, for example, is a text on planning:

But the impact of uncertainty is not always a cost; it can occasionally actually improve an outcome. Administrative action is not simply a matter of deciding what to do. Decisions often result from lengthy processes that involve all sorts of interpersonal wrangles and Machiavellian tactics. Uncertainty can help to bring these processes to the point where some measure of agreement is achieved. Given the ambiguities and cross-pulls of political life, given honest differences in values and factual judgments, I wonder how often people would agree on a course of action if everyone knew precisely what they were agreeing on. The uncertainty inherent in all aspects of [a] decision can provide the leeway for a rearrangement of fact and emphasis which makes coalition possible and a strategy of achieving consensus effective. The uncertain world must fight fire with fire. In this sense, the obfuscation of uncertainty can be an advantage rather that a cost.[11]

This is not just the case early on in the design process. Uncertainty, the unknown, and ambiguity are evident all along the route. In retrospect, at any point in the process, the past can always be made to look clear, even when it offers evidence of missteps. Indeed, to label a piece of history a *misstep* requires participants' assent. Missteps, too, are socially construed.

As you approach the here and now, the more diffuse, unclear, uncertain, and ambiguous do events appear. The further back you go in time, the less diffuse, the more clear, certain, and unambiguous they seem. Company lore lies back there somewhere. Being there, at any stage in the process, one sees the uncertainty, the ambiguity, and all the different perspectives participants have on the design task. These are continually in motion, being redefined, reconstructed, made reasonable.

While agreement and consensus are essential to moving forward in the design process, it does not necessarily follow that all participants share the same vision of the design even when consensus is achieved and even when the artifact has been realized in hardware. To see this, we turn next to a meeting held in the office of the president of the firm building the x-ray inspection system.

Precollimator #3

This meeting was called to decide if the design should be amended to include a third collimator. And indeed, in contrast to some of the meetings I have earlier reported on, this decision was made. Participants gathered fifteen minutes after the scheduled hour and took seats around a rectangular table located in the middle of the president's office. One end of the table stood a few feet from a wall, upon which a blackboard was hung. The president sat at the opposite end, facing the board. Behind him, across the room, lay a broad expanse of windows. Seated on the two long sides of the table were the chief engineer, the project manager, the project scientist, and the head of electronics.

Two questions were raised at the outset of the meeting, after the usual settling in. The president asked, "Will it help?" This prompted a discussion about the physics of the instrument and how the image might be improved if the precollimator was included. The question was specialized to "How much scatter will be eliminated?" The project scientist claimed that it would cut scattered radiation by a significant factor. He had already done an analysis, based on a simple mathematical-physical model. The president, himself a physicist, had constructed his own model and offered up a different scenario; he appropriated the question on the spot, making it his own in seeking to understand the project scientist's claim. The project scientist, at the board, outlined his analysis, responding to the President's questions and commenting on his boss's model in turn. This discourse was a two-person event between the president and the project scientist.

A second question engaged the others at the table. The financial officer had asked, "Do we really need it?" This stimulated a second dialogue—literally cross talk between the chief engineer and the project manager, a dialogue that went on simultaneously with the project scientist–president exchange. Here, though, the concern was cost—whether the addition of precollimator #3 would be worth it. This discussion had a global orientation, addressing the system as a whole.

The chief engineer thought of the precollimator as "insurance." He saw a risk, albeit small: "If we don't go ahead and build it, add it to the system, the system won't perform as well as we want." The cost, while not insignificant, did not appear to any of the parties as major. Still, time, resources, and people would have to be committed to design and fabricate the item. The independence of the subsystem was also noted; adding this new piece of hardware would not impact on the design of other subsystems. It is a passive device; as long as there was space for it in the test chamber, its design could proceed independently.

The two dialogues, sometimes independently, sometimes intersecting, and sometimes interrupted by a foreign topic (Can we inspect a small car with the prototype system?) went on for a half hour or so. At the end, everyone agreed to add precollimator #3.

The two conversations show two different perspectives on, and express two different interests in, precollimator #3. The project scientist and the president see the "physics" of the device; for them, "improvement in performance" means a better-quality image, and they spend their time and energy in mathematical modeling, testing and jousting with its adequacy, trying alternative models to see if they might give further insight into the "physics of the situation." The sense I had listening to this dialogue was that if they developed full confidence in their abstract models and their mathematics, from which they had deduced that precollimator #3 would indeed significantly improve the image, there would be no question but to go ahead and build the thing. Cost was rarely mentioned; it was there, but only in the background.

In the other conversation the cost was central, not as a number but as a major factor. Here the language was that of accountancy, not physics. The metaphor was actuarial; the story was about the risk of system problems, about the insurance the hardware would provide, about the resources required to build the collimator, and about other global concerns.

Both parties were aware of the other's conversation and concern; and although they hit the same note on occasion, the two remained disconnected to the end when they decided to add precollimator #3.

The ending was not a synthesis of the two discourses. One party concluded that the addition would significantly reduce scatter, the other that it could be designed and fabricated at a reasonable cost without undue pressure on anyone's schedule. The two were consonant, but there was no instrumental method applied to bring the two perspectives into play. The trade-off, if we can even call it such, was weakly constructed since the marginal cost appeared small relative to the overall budget while the reduction in scatter appeared to be significant to the physicists as measured by percentage reduction. But, I repeat, there was no valuation of marginal cost of marginal improvement of the image in adding precollimator #3. The design decision in this instance is best seen as an overlay of different interests rather than a synthesis within some flat, cognitive domain.

What is precollimator #3? Precollimator #3 is our insurance policy. Precollimator #3 is two parallel columns of lead plates held in position by a stable and rigid steel structure accurately positioned in the path of the beam. Precollimator #3 is an absorbing boundary in a physicist's model and sketch on the blackboard. It is all of these.

The meeting in retrospect had a perfunctory quality. The decision was not difficult. It was prefigured by the work of the different participants within their own object worlds. The project scientist came to the meeting with a model of how precollimator #3 would reduce scatter, and by how much. He had had the mechanical engineer on the design team draw up some sketches of a structure that would support the heavy lead plates, whose probable dimensions he had calculated. He himself had roughed out some estimates of costs, thought about contracting out the fabrication, and talked to the shop to see what their time commitments were like. The president had thought about how they could fit a small car through the prototype inspection system. Would the presence of precollimator #3 interfere at all with that possibility?

In the course of these developments, participants had made cost estimates, developed analyses of performance, sketched a mechanical structure, checked out a fastener sent by a supplier, and redrawn the plan view of the test chamber with the footprint of precollimator #3 shown. Production had even noted, on their schedule, the possibility of a "PC#3 Job." All of these bits and pieces are the material manifestation of the design of precollimator #3. The form, as well as the focus, of all of these bits and pieces of design activity prior to the meeting vary according to the different responsibilities of participants.

They are transient. They are plastic, subject to change. And they are continually being negotiated as participants move the design forward to its realization in a functioning artifact. The decision to go ahead with PC#3 was one step along the way.

In time—with further detailing, with further negotiations both inside and outside the firm (the project manager contracts out a good part of the fabrication), with the accumulation of more notes and sketches, and then with formal drawings, purchase orders, real cost estimates, and real hours charged to this subsystem—participants move the design of PC#3 to completion.

Parts arrive. Jim makes space in the lab. People in the shop lay out the fixtures for assembly of the parallel lead plates. The project scientist stops by now and then to check on progress, in particular whether, in assembly, tolerances on the positioning of the lead plates are being maintained. The support structure is tightened up, and PC#3 is hoisted upright. A last-minute fix is made to counterweight the structure to ensure its stability. PC#3 as artifact now exists.

Bills are eventually paid. Flags are removed from production control charts. Sketches are thrown away, and computer files of input to a structural analysis program are erased. A summary of the physicist's model is pasted in a lab book. Formal drawings are coded and stored. The artifact is dismantled and boxed for shipment to the West Coast.

7

Endings

Solaray

It is August and it is hot in central Massachusetts. The haze and humidity, while reducing the solar radiation falling upon Mrs. Jones's rooftop photovoltaic array, only conjoin to intensify the heat. By midday Mrs. Jones had resigned herself to the weather but she remained disturbed, questioning the workings of her high-tech energy system. She had enough experience with the system, designed to provide a significant percentage of her household's energy needs, to know that something was wrong. Twice this morning she had been down to the basement to check the scale of red indicator lights on the front cover of the power conditioning unit. Only one light of ten was lit. Yet she knew that, on a day such as this, just an hour past the time when the sun reached its full height in the sky, the array ought to be producing 80 percent of full power. Eight of the ten lights ought to be glowing.

She checked her electric meter in the garage. From the rapid rate at which the meter's disc was rotating, she surmised that the utility company was supplying most of the power to the television, the refrigerator, and the clothes dryer that she had just, reluctantly, switched on. If the array was working as it was designed to work, she expected to see the disc rotating slowly if at all; indeed, on some bright clear days, without the dryer engaged, she had seen the disc rotating backward, indicating that she was supplying power—the excess produced by her photovoltaic array—back to the utility!

She called one of the numbers Solaray had left with her for occasions like this. Tom asked her a few questions: Did she have power? Were any of the indicator lights on? Was the horizontal, flat disc, barely visible through the glass front of her electric meter, rotating? In a half hour he called back, told her not to worry, that indeed something was awry, and that he would be out the next morning to diagnose the problem.

Tom was having a good week that week until Mrs. Jones called. He had just received word back from Sacramento Power on the West Coast that the assembly and installation of a large field array made up of Solaray's product—a big system rated at 200 kilowatts—was proceeding on schedule. While the basic power-producing unit was the same res module, eight of which covered most of Mrs. Jones's south-facing roof, Sacramento Power's array required one hundred times this number to produce its rated output. A preliminary testing of a subarray indicated that the large-scale system would perform as efficiently as designers of the residential module had planned. And all who saw the field array under construction admired its appearance. Here was a major step forward for Solaray, a significant "scale-up" in the firm's productive activity.

It didn't take Tom long to discover the fault in Mrs. Jones's system: There was nothing wrong with the array or with the power conditioning unit (the "inverter"). A diode had burned out in the string combiner box, the enclosure within which all of the leads coming down from the rooftop modules were connected together to give the high voltage required at the input to the inverter. It would take a few days but Tom would fix it himself back at the lab; he would replace the diode, add an additional heat sink, then return to install the box back under the inverter in Mrs. Jones's basement. He assured her that in two days she would have her electric meter running backward again.

News from the West Coast about the Sacramento installation generated a good bit of excitement throughout the firm. Yet while the res module appeared to be launched in good form and on its way in the world, design activity continued: There was a new frame structure on the drafting boards and another, less costly option for the glass superstrate under consideration. Systems Engineering and Panel Fabrication had generated alternative network designs that made possible other nominal module voltages without excessive contortions of the 48-volt design. But more intriguing was the rumor making the rounds at the firm that Cell Production was going to make a major commitment to a new crystal growing and processing procedure, one that had been talked about for the past six months but until recently had been considered too risky to pursue. The goal was to reduce the costs of production of the photovoltaic cell, and hence module costs, to a point where Solaray's technology would become truly competitive as a power-generating option within the electric utility sector.

Then, the following Monday, Tom and his colleagues in Systems Engineering received the word: A major decision had been made in that past week by corporate headquarters in New York. All effort was to go into the development of cell production technology. Module production for direct sales was to be cut back to a minimum. All module design tasks were to be terminated. All effort was to go into the design and development of cell processing and production technology. The Systems Engineering and Marketing Group was to be reduced to one person. Tom was to move over to Cell Production; the rest of the group received notice of termination of employment.

I started this book with a question: What fixes the form of technology in our lives? More specifically: How goes the process of design within today's engineering firm? I raised the question of autonomy and asked how ordinary citizens might have more of a say in the process. I now want to summarize this book's response.

Initially, I set up two straw people: One had the theories and methods of science determining form; the other relied on market forces to select among alternatives. I argued that, while science and markets have roles to play in design, they are far from decisive.

For example, science matters in that it provides the underlying form of designs—the physics of precollimator #3 at Amxray, the chemistry of Fritz's urge to understand the dropout problem at Photoquik, and the rules that define the performance of a photovoltaic cell at Solaray. And a single field of science may predominate within each particular subculture. We can even usefully speak of a founding science or paradigm as the source of innovation and entrepreneurial activity at each of the three firms: solid state physics for photovoltaic material; radiation and electromagnetic field theory for x-ray inspection machinery; chemistry for the technology of photoprint processing. Science, in a more general sense, is the mode of thinking within object worlds. It also structures the way in which participants frame their work process and interactions. The strongest claim that one might make is that science, in this socializing sense, controls design process. But this is not the science of those who hold that science determines the form of technology.

The scenario about science determining form, as ordinarily understood, misses the complexities of alternative forms and paths to a design. It ignores the diverse interests of participants in the design process, each making claims based upon scientific rationality, and it fails to acknowledge the indeterminacy of technical constraints and specifications and ignores their negotiation in process.[1]

As for markets, we do hear participants use a market vocabulary of efficiency and profits in their designing. They talk about costs, demand, and customer needs. Recall the meeting at Solaray where Ed and Tom constructed a map of the order/shipment process or Sergio's strategy for solving the dropout problem at Photoquik in terms of meeting the needs of his so-called customers, that is, the group working on Atlas and another product team out in Colorado. The definition of *explosive* as a device with wires can be viewed as acknowledgment of the needs and interests of the users of Amxray's product—both the radiologist at the computer console and the terrorist checking his or her baggage.

But while this vocabulary suggests a concern with the world beyond the firm, in most of these instances the attention to user needs and markets has an ad hoc character. Statements sound like rationalizations meant to buttress a proposal, and cost appears as a chip to wager in negotiations, used in a claim—rarely substantiated—intended to support a favored design alternative and deny another.

The prophecies of market studies, while given little credence by participants before the design is realized, are used in the same way. Toiling within different object worlds, participants cite them to justify their own individual proposals. The market, then, as construed and used in the process of design, is available to justify alternative forms. How can markets be determinate if this is the case? On the other hand, this scenario, in contrast to the one that relies on the dictates of science, presumes too much flexibility in the design process; it misses the constraining influences of social norms, technical constraints, and infrastructural limitations.

Having committed our straw people to the bonfire, we go inside the firm to study how choices are really made. There we see designing, in part, as work within the object worlds of different participants. We sit beside Beth at Solaray as she negotiates with her computer, reworking her model of a stand-alone photovoltaic-powered desalination system for Saudi Arabia. We listen while Arnie and Jim at Amxray probe and exercise the tentative machinery of their data acquisition subsystem in search of the source of a faulty byte in memory. We watch as Don works as hard as he can to size an air tank to fit into too small a space on the new Atlas photoprint processor. We don't make a sound or dare interrupt.

But work within object worlds is only one piece of designing. Contemporary technology is an intersection of multiple object worlds; its design cannot be split apart into a collection of separate tasks independently pursued; it requires instead the continual engagement of, and exchange among, individuals schooled and trained in a range of disciplines.

The object is not one thing to all participants. Each individual's perspective and interests are rooted in his or her special expertise and responsibilities. Designing is a process of bringing coherence to these perspectives and interests, fixing them in the artifact. Participants work to bring their efforts into harmony through negotiation.

This harmony, or lack of it, will be reflected in the artifact or in the "built form" (a phrase used by architects that fits nicely here). The quality of the final design and artifact, as evidenced by the harmony of the different underlying forms of different object worlds achieved, will then depend upon the social process engaged by participants, the competence of participants working within object worlds, and also the infrastructure and its vital, sustaining ecology. A major claim of this study is that this process transcends rational, instrumental process.[2]

If someone asks, "What is the design?" at any point in the process, I respond, following Durkheim, that it exists only in a collective sense. In process, the design is not contained in the totality of formal documentation, nor is it in the possession of any one individual to describe or completely define, although every participant will tell you his or her story if asked. This is the strong sense of "design is a social process."

Designing is not simply a matter of trade-offs, of instrumental, rational weighing of interests against each other, a process of measuring alternatives and options against some given performance conditions. Nothing is sacred, not even performance specifications, for these, too, are negotiated, changed, or even thrown out altogether, while those that matter are embellished and made rigid with time as design proceeds. They themselves are artifacts of design. So, too, with other constraints; even codes have to be given a reading and an interpretation. They are all there to be negotiated if those readings run in conflict. Specifications become artifacts of process, reconstrued in the engaging of different perspectives of different object worlds.

I do not mean that constraints are fully a product of the social imagination or that their technical import is wholly a matter of the immediate context, a product of local consensus. But they are not as definitive as they first appear (or later appear in retrospect). An instrumental, object-world reading out of context does not capture their true meaning or full significance in designing.

This same shallow reading of the instrumental methods employed in designing—Critical Path Method, Milestone Chart, Pugh Method—likewise misses their true meaning and value in process, namely, how they function to provide a framework for negotiation.

Consider Solaray's block diagram for the order/shipment process. Its value lies not in itself as an instrument to guide the order/shipment process, but in the social consensus achieved in its construction. It stands as a contract summarizing an agreement. It crystallizes, in an agreed-upon pattern, understandings about relations among individuals, though the individuals are not explicitly identified. It is itself an artifact of design, and it also hides the important process whereby it was constructed.

So with the other methods—designing is also going on in their making. That is what is most significant. The final chart is hardly interesting and rarely referred to, unless it later shows a bug or is challenged by further developments. But if that happens, the negotiation process starts anew.

Instrumental thinking within object worlds shapes participants' visions of all that goes on within the firm. We have seen how the object influences the organization of design, how participants read into their day-to-day workings and negotiations norms derived from object-world paradigms and thereby attempt to control the uncertainty and the ambiguity of social process. We have seen how this is reflected in the humor pasted on the walls (truer laws of process).

The object prefigures participants' expectations and rewards, as evidenced in the aesthetic value accorded efficiency or how the immediate experience of the thing working, deterministically, according to design is celebrated. The object in this way is its own reward.

The object is certain, determined, abstract, conserved, and so on. The process of designing is ambiguous and uncertain; there is the unknown. Ambiguity and uncertainty are especially evident at the interfaces where participants from different object worlds must meet, agree, and harmonize their proposals and concerns. Ambiguity serves them well in this regard. It allows them room to maneuver, to reshape, to relearn and come together again.

Uncertainty and the unknown are also to be valued, for they make designing the challenge that it is. A disaster meeting is not a disaster. The struggle over criteria to judge design alternatives for the fix of the photoprint processor is both a developing of shared meaning of terms and designing in the sense of foreshadowing and framing what the artifact will be. The two are inexorably mixed.

The event at Solaray is a disaster of another kind; it is not contained within this summary; it is a disjunction of another order, a gap more severe than that among object worlds. It suggests the need for new study of another grouping, of managers and corporate chiefs, and how they "design" within their world. The connections, exchanges, and relationships between corporate strategists and participants in engineering design appear especially worthy of investigation.[3] The event, for this study, marks closure of the module design process at Solaray, an ending of a cataclysmic sort—not of the kind I anticipated when first drove west, but there it is.

Photoquik

Sergio, reading between the lines of the memorandum from his chief (MH) was ill at ease. He did not look forward to their upcoming meeting. It had been six months since he and Hans had narrowed down to three the number of options originally considered as possible solutions to Atlas's dropout problem. Some progress had been made on all three; while not out of the woods by any means, they had a plan for bringing at least one of them to fruition. On some of their better days they could catch a glimpse of an open field ahead.

Now the ground was shifting again. Sergio had already learned from Hector that Atlas wanted Hans moved over to work with them. Atlas was in a panic mode—the quick fix they had developed was showing some erratic performance. They wanted Hans with his chemistry expertise on board full-time.

It didn't take long for MH to get to the point: Hans, Phil, and another tech were to move over to Atlas for what little time remained prior to launch of the product. Sergio was to continue with the two mechanical options—the lead-in paper-path change and the electromagnetic excitation. Atlas would go ahead, without his help, perfecting their quick fix. Sergio's effort was now limited to longer-term objectives—providing Leonard in Colorado and future products with a convincing solution to the dropout problem. As MH explained, the pressure was off.

Sergio objected. Hans and his technicians were doing important work for him. Their predevelopment treatment studies were beginning to show consistent and promising results, and Hans was also invaluable as the one chemistry person he could call upon when the E&M or the paper-path lead-in efforts ran into difficulties that seemed to be chemistry related. And what about his schedule? He had set Taguchi testing for October; that testing was to be the first meaningful evaluation of all three options. His plan had been to develop and test the three options in parallel, then go back to Atlas and Colorado with the results. His whole approach now was challenged if not reduced to naught. He would no longer have adequate resources to proceed as he had planned.

MH listened but wouldn't budge. Atlas was top priority. A product program at this stage and in these straits was not to be denied.

Back at his office, Sergio met with Don and Hector to give them the news. He instructed them to redraft a schedule of work still to be performed that reflected the departure of Hans. The testing set for October would be rescheduled for some time in the future. Sergio would be taking the next two weeks off.

Sergio's world had collapsed. The formal memo he received shortly thereafter read rationally enough—a bit of a cutback in one phase of the effort, a shift of several of his people to Atlas, while work was to proceed with two of the three options and he was assured of the resources for these two remaining tasks—but this reasoning did not truly reflect his diminished role.

What had been lost? Sergio lost a key person in Hans. He had been forced to cut out one of the technical options he valued. His whole approach—"We will present you with a menu of options. We will tell you how much they will each help you eliminate dropout, individually or working together. We will tell you how much they will cost. You then decide what you are willing to pay for the performance you desire"—lost its force. In effect, Sergio had been put on hold, and to be put on hold in the business of engineering design is to be cut out of the action. It didn't look that way on paper; the words of the memo did not appear so meager, but it was indeed the end of Sergio's designing a solution to the problem of dropout. The problem as Sergio (and his team) had construed it was over.

Designing, once again, is done in contexts—settings for the playing out of different individual and collective interests. The setting can be a lone engineer working within his or her own object world carrying on a dialogue with the object, negotiating alternative representations of underlying form; it can be the world of the design team engaged in a social process of negotiation and collaboration at the same time; or the setting might be the whole firm, or beyond. Here, in Sergio's experience, is a specific instance of how organization and the definition of context for a design task legitimate, indeed form the essence of, that task. For Sergio's design task is a design task only as long as he can sustain the context he has worked so hard to establish, only as long as the dropout problem is defined in his terms, as a serious problem, because (we agree) it can significantly degrade performance, because (we agree) other product lines have had, and will continue to have, the same problem shown by Atlas, and because (we agree) there are multiple solutions and, although we don't have full knowledge of its cause, we presume the same source accounts for all evidence of dropout seen to date. MH agrees (we observe) to provide the resources—people, machinery, funding—to solve the problem. Through these agreements, Sergio constructs the dropout problem itself as well as three solutions to pursue.

We see now, and not until now, the fragility of Sergio's world and how hard he had worked to maintain the problem as a problem to be accorded its due. And once more we see how corporate decisions dramatically alter the trajectory of engineering design.

Contexts for designing are shaped, constructed, maintained, and destroyed. To understand design process as it is, one must accept this context making and unmaking as part of the process. It is not enough to focus on work within object worlds, to speak of negotiation of trade-offs within some well-defined boundaries if our aim is to understand the full complexity of designing as social process. That process must include the invention and elaboration—the design if you like—of the milieu itself within which the participants work. At Solaray, and now at Photoquik, the destruction of context makes this clear.

Amxray

It's the beginning of fall in Boston and time at Amxray to pack up the hardware, including precollimator #3, for shipment to Palo Alto, where there is a sufficiently powerful x-ray source and a large enough test chamber to accommodate all of Amxray's apparatus and enable full-scale testing of the prototype.

I'm in Michael's office listening while Jim relates the latest bit of folly that often intrudes in design. It seems the truck driver contracted to make the haul west has proposed to take his wife along on the journey. This in itself is not objectionable, but Amxray wants one of its own technicians to ride shotgun, in order to keep tabs on the condition of the hardware, the only cargo on board the mammoth trailer. Amxray's representative is to check in with Michael every day. After a meeting and a few phone calls it has been decided that the driver's wife can go along; three people will make the trip, but the itinerary has been chosen to accommodate Amxray's envoy.

Mark, the team's software engineer, enters the office looking disturbed:

We've got a problem. Our data processing is taking too long. There may be a bug in—

Michael intervenes:

Michael *How much time? What do you mean? We had it working smoothly last week . . . the A-to-D read cycle. Is that it?*

Mark *No, we had budgeted for that. No, it's not that so much . . . in the data analysis blocks . . . not easy to pin down.*

Michael *Yeah, but how much time . . . how far off are we?*

Mark *Close to two minutes. It's taking over two minutes.*

Michael *Two minutes! That's a helluva long time to get one vehicle through. I can see them lined up now. How are we going to meet our productivity spec? Might as well go on board and open things up! What are we supposed to be at? We were aiming for under thirty seconds, no?*

Mark *. . . think I can get it down . . .*

Michael *Where is the problem? Which block? Two minutes can be a very long time sitting, waiting in front of a computer.*

Mark *I don't know yet. It looks like a bunch of little things. I bet a good bit of it is in the lines of coding we have in there for ease of debugging. All that out should gain us at least a minute.*

Michael *You know we're all ready to pack up. Next Monday is the day. How are you going to work if the computer is in a crate? Do you have another platform you can work on?*

Mark *We have a backup, but I'd rather not leave the prototype. I don't know if it's configured the same way. [Arnie] was using it last week. Let me take a crack at it this weekend before you pack up. You are not really crating the console until Monday afternoon, right?*

Michael *Think you can work it through in a day or two? What will you need? Want me to come in? What can we do? Got to get it down.*

Mark *Just make sure the heat in the building is left on. I don't need help. Everything is on the hard drive and I've got it backed up . . . plenty of test signals. Let me go to it alone.*

Mark spent fifteen hours of a beautiful, crisp fall weekend sitting at a computer console, wrapped up in an object world. He methodically went through every block of computer code, timing each phase and segment, looking for unnecessary operations—the debugging steps were the first to be eliminated—and for ways to restructure segments in order to achieve the sorely needed time savings. Life is intense within this world of bits and bytes, where a characteristic event such as the execution of one line of machine code takes a millionth of a second, where data bytes from 512 detectors, sampled at thousandths of a second, mount up in a queue to be processed, and where the code that does the processing marches through its programmed operations millions of times to yield the correct grey tone of each and every one of a thousand pixels that, all together, distributed over the monitor screen, constitute an image. All of this activity in the production of an image has to take less than thirty seconds if the performance specification is to be met.

Mark has specific tools at hand for his diagnostic work—software development programs that allow him to step through code line by line or run through a subset of instructions over and over again in order to obtain a measure of the "real time" the subset consumes. Using these, he develops a feel for the machine's response time at the level of microseconds. With these tools, he appropriates the whole of the program in all of its blocks and stages. He alters a line of code, or eliminates a section altogether, runs the program, and notes the difference in execution time. He inventories microseconds; they absorb all of his attention as he tracks down and conserves these intervals as objects. Backing and filling, he labors to reduce the overall time to produce an image. Alone in the building, he hardly notices the hum of the heating system as the time flies by.

On Monday morning Mark wasn't at his desk, but there was a note on Michael's: "Got it down to twenty seconds . . . think we can live with that. I'm getting some sleep but will see you late this afternoon." Michael felt a large weight lifted from his shoulders. He could pack up for the West Coast with confidence that the system would perform as he had hoped and fully expected.

There were no incidents in transit. Michael, Mark, Arnie, and Jim flew out the day the truck arrived. Setup in the test chamber went well. The cabling proved long enough (there had been some concern that it would not); the drive track over which the containers would be pulled through the path of the beam worked flawlessly; precollimator #3's placement proved no problem; and in two days, Arnie had a decent image up on the screen.

There is a craft in this, too. The intensity of the light level of each pixel on the screen can be varied over an essentially continuous range of greys from white to black. There is the opportunity to do "contrast stretching"—to accentuate or depreciate the contrast between two signal levels as picked up at the detectors. There is a filtering process Arnie can "call up" to give edge enhancement: an algorithm that massages the relative intensity of the pixel values in a small neighborhood and heightens the contrast of what it is programmed to judge as a possible edge. And there are zooming and scanning capabilities as well.

Arnie is, with all this freedom, practicing a new high-tech art form. Few others can compose an image the way he can. He knows the inner workings, the underlying form, of the filtering algorithm. He knows what visual effects are obtained by manipulating the grey scale or through contrast stretching, and he is an accomplished performer on the joystick.

Like Mark, on his object-world weekend of code fixing, Arnie is intimately involved with technique. In this object world, mind and hand must work together, but now the "handling" is highly visual while linked to the play of joy stick, which, in turn, is tuned to the inertia of the coding within the machine. Arnie knows this craft better than anyone else.

When the project director from BG arrived the next day, he was pleased with what he saw. Arnie had stored on disk the data from several trials of the system made during the past few days. He put on an impressive show.

The project director brought with him his marketing director and asked Arnie to instruct him in the manipulation of images. BG's Marketing Division was to take full responsibility for demonstrations to potential customers, relations with the press, and so on. But, although Harold learned quickly, Arnie was uneasy during the first set of demonstrations. It took Harold too much time to get the best image, the proper contrast setting, the right framing. Sometimes he muffed it, lost the image, lost control of his massaging, and simply went on to the next run. Needless to say, Arnie was on edge. He knew he could do better.

Still the demonstration was judged to be a success by all interested parties—management, engineering, marketing—as evidenced by a San Francisco Bay Area newspaper's front page report, with photo, proclaiming the virtues of this contraband detection system. The artifact had become an entity in its own right, slipping out from the grasp of Arnie, Mark, and Michael. Like a child no longer a child, but off on its own for the first time out into the world, shaped and molded to the interests of others, its behavior in its stumbling first encounters off home turf

surprises you. It was no longer Jim's, Arnie's, or Mark's alone to appropriate, manage, and know. That night they gathered for Mexican food
at a local restaurant and watched the Red Sox lose a critical game of
the play-offs.

This was more of a storybook ending for the design task than the
events at Solaray and Photoquik. Closure is displayed in two respects:
First, we have a "handoff" of the artifact to the world outside the firm;
second, we have a dispersal of the design team. Jim, Mark, and Arnie
will no longer be working together as a team when they return to the
East Coast. Michael still has some tidying up of loose ends and work to
put in on the proposal for the full-scale production system, but, like his
teammates, he will soon go home to take on new design tasks.

Other Endings

A vitalist's metaphor: When designing is in process, that process is alive.
The object is alive and laden with uncertainty and ambiguity. That is
what makes designing the challenge it is. When the design is complete,
when the product gets shipped, when the documentation is printed,
and most significantly when the team disbands, then the process is over.
The object as artifact is dead within the world of the firm; it no longer
serves as the occasion for surprise. All is in order; all functions
deterministically.

No wonder most documentation, in its description of what was once
alive as a design project, reads like an obituary: Writing about the object
as a fully defined artifact for outsiders is often a painful task, a reflection
of the inadequacy of object-world voice and the written text to capture
the object as social process both in its design and in its performance
out in the world. Only a few take to this finalizing design task.

Still other lives are probable. Other cultures or consumers may appropriate the artifact and make it their object. There are other stories,
other social processes of impacts, of alienation, reconstruction, and use.
The artifact as object can live again. It can become a nexus or icon of
social discourse and exchange. In its use it can impose, block, enable,
shape social connections and the aspirations of those it meets. There
are other object worlds within which the artifact can be seen and used
in different ways. Deconstruction and bricolage are always possible.

Some day it may sit in a museum, repainted to deceive visitors and to
reconfigure the past, lifeless no matter how interactive the exhibit. Or
it may join its cousins in more mundane settings—a junk pile behind

the barn, toys in the attic, there to come alive briefly in the hands of some child's fantasy. Some limbs still move and inspire. All of which, whether it does or does not, is a matter of the possibility for social exchange.

Technology is object and technique, but object and technique *inside* culture, not external to you and me. It is integral, constitutive of a seamless web,[4] but transcending science and its logic, political power, high finance, global infrastructures and markets. It weaves through our everyday thinking, educating, family raising, churchgoing, leisure, and labor. Technology as artifact, as system, tool, productivity, efficiency—yes; technology as metaphor, as process, as values we live by and in as well.

All of this about closure and the endings of design is part of the "What is technology?" question. The difficulties we have in making a response, the unease and insufficiency we experience reading histories of technology, the sense we feel when facing an artifact in a museum—with a barrier between you and it that reinforces the independence of the artifact but curiously also says at the same time "I was once alive"—all of this unease and confusion are due to the deceptive "hardness" of the artifact. We see only its Cartesian structure.[5] The stone my foot kicks is out there.

The point of this book is that there is more there, or rather here. The objective reality of the technological artifact is a social construct, as much a text to be read and interpreted, as much a pattern of behaving to be appropriated and shaped to one's own use and pleasure, as much an expression of history and values, as it is a stone to be kicked or thrown. Different users can appropriate the artifact differently as object; Robert Pirsig even goes so far as to suggest that the artifact has a personality, constructed conjointly with its user.[6]

We come to a critical juncture. Are there better ways to do designing? Of course there are alternatives. Only after the artifact is fixed does it appear otherwise, as rational, with no other way to meet the market niche, no other shape that would embody its internal logic. In process there are many objects, many potential artifacts, many object worlds.

I posit an axiom: There are always significantly different design alternatives given the same initial conditions—performance specifications, resources, infrastructure, and the like. We have seen enough to throw out the hard autonomy argument—the argument claiming that instrumental factors alone account for the design. Science push and market pull, optimization and satisficing are not determinate.

Can there be better designs? Of course. If design is not autonomous, if significant alternatives are possible, then some will be better than others. The question itself, once one accepts the axiom, is not very interesting or challenging, even a sort of tautology. The really important and interesting question becomes: What do we mean by a better design? More important still: Better for whom?

This is a significant displacement: To shift attention from the artifact to the consumer or to its designers—any social context of its existence in the making or in use—opens the door to the possibility of constructing meaningful responses to questions about quality. Otherwise, with our gaze frozen on the fixed artifact, constrained to speak in instrumental and object-world terms, only sterile stories are possible. The significance of the diversity of interests of parties and peoples in judging quality is key.[7]

Tacking a bit in object-world terms, we might think of two dimensions girding "quality" and "better": One is the "who," the set of interested parties to the artifact or system, the other is the set of criteria by which these parties do their judging.

The parties to this valuative process are diverse; they go beyond the participants in design to include the state, the community, individual consumers or users, and even their neighbors, and we ought also to include those who provide service as distributor, as salesperson, as inspector, or as repairer. Even when we restrict our immediate attention to participants in design, we know enough now to distinguish among individuals working within different object worlds, and in the light of events reported earlier in this chapter, we might accord the managerial class a special status.

The criteria these different people, agents, and institutions will employ in judging quality will be as diverse as they are: For some, sales and profit margins will be the most important measures of quality; for others, efficiency will be primary, or robustness of the design, or ease of maintenance, or safety features. Some will value an artifact or system if they can open it up, appropriate its workings at a more fundamental level and alter it to their own purpose; others, including the user's neighbor, might want only not to be bothered. Low cost, of course, will be treasured by most; high cost by those who seek to display their artifact as ornament.

Responsible parties might ask: Does it recycle? What stream of wastes are generated in its development, production, and use? Does it mean jobs for the community? Does it mean *good* jobs for the community?

(Again *good* must be socially constructed.) How will it fare in the global marketplace? Does it contribute positively to our balance of payments? Does it help build our infrastructure?

There are, then, many different parties, different interests and criteria for judging quality. The criteria are not independent. They intersect in the object in particular contexts. So, too, will the parties to evaluation overlap and intersect in the object within particular contexts.[8] The point is that it is meaningless to talk about quality in terms of the artifact alone. Context, object, parties and their interests—all must be attended to.

But how do we achieve better designs? To respond requires more than a recognition of the diversity of interests and criteria that might be used in judging a design. The question becomes: How can these parties, each with legitimate interests, affect the process of design?

We have seen that some parties have direct and immediate control over the design process. The financial success of a venture is going to figure largely in decision making; but even here there is uncertainty and an occasion for negotiation, at least among some parties to the process. Then, too, estimated returns on an investment will be constructed through a discourse about market expectations and customer needs— our marketplace-oriented straw person does play a role.

The technical constraints of the object worlds of engineers give them considerable authority in design. But science and instrumental reasoning alone do not fix quality: Constraints are negotiable, instrumental method is an occasion, and quality alternatives are possible.

History matters. No design begins with a absolutely clean sheet of paper. Last year's model, a competitor's product, the availability of new tooling, new materials, new information-processing machinery, the recently enacted R&D investment tax credit, the current recession and the consequent weakening of the infrastructure, a free-trade agreement and lower labor costs—product and system design, quality and better designs, are situated within particular historical, political, and cultural contexts.

Here, then, is the full complexity one confronts in search of quality and better designs. Yet it does not follow that criteria for better objects in diverse contexts cannot be formulated and put into play by a wide range of people with an interest in doing better, indeed, with the responsibility to participate in shaping our technological future. What does follow is that the formulation and implementation of criteria must be a social process of negotiation.[9]

In those negotiations we should not presume that there will be one design that is best for all or that all criteria ought to carry the same weight. On the other hand, we should not be discouraged. It is time to dismiss the sort of fatalism that Jacques Ellul and other determinists have preached; we cannot give up so easily. It is time to grow beyond the myths and fantasies of well-meaning but naive technocrats; we cannot leave the job of design to them alone. Autonomous technology is an empty metaphor. What we can insist is that the process be a quality process, made open to a full range of people and institutions, organized and orchestrated to promote a free exchange of their legitimate interests in better designs.

We end, then, where we started, with questions like: Do you know how your telephone works? Or, how can we build a better telephone? But our inability to respond in as explicit a way as some might desire does not imply that we have gotten nowhere. To the contrary, the realization that design is a social process, that alternative designs are possible, and that a design's quality is as much a question of culture and context as it is of a thing in itself or of the dictates of science or market forces—all this is prerequisite to moving beyond simplistic images and myths about technology and doing better as designers, as corporate strategists, as government regulators, as consumers, and as citizens.

Notes

Chapter 1

1. The names of the firms and the participants have been changed in accord with the usual norms on privacy. In some cases, the technology has also been changed to protect its identity.

2. "Only 19 percent believe that they have a clear understanding of how their telephone works." From Jon D. Miller, "Technological Literacy: Some Concepts and Measures," Public Opinion Laboratory, Northern Illinois University, 19 February 1986. Note the inferred wording of the question. In this form, it is as much a test of a consumer's confidence in his or her technical understanding as it is a test of the latter. The qualifier "clear" no doubt assures a low positive response. Evidently, we will never know what respondents do know about how their telephone works. But note that, with the question phrased in this way, it is not necessary for the sociologist to know the slightest thing about how a telephone works in order to complete his research. I assume that a graduate student did the dialing.

3. This is not to claim that "anything goes"—that is, that some respondents will not offer erroneous accounts in their attempts to please a sociologist. Nor should one conclude that the telephone can be put to use for any arbitrary purpose or that the telephone is value free or neutral. It is a long way from the claim that different persons have different relationships with one and the same technology, the claim made here, to the fantasy that depicts technique as the sterile embodiment of nature's laws, devoid of human intent and interests.

4. The exclusion of ludicrous but rational and logically possible readings of statements and claims from the serious business of design is a matter of context. The neophyte engineer's suggestion will often be greeted with a laugh, and he or she learns, in this way, that "a horse is not a chair around here."

5. Christopher Dresser, *Principles of Decorative Design* (London; New York: Cassell, Peter & Galpin, 1873), quoted in F. Russell, P. Garner, and J. Read, *A Century of Chair Design* (Chicago: Academy Editions, 1980).

6. The knowledge that three points of support are all that are required to support a seat (and to define a plane) is reflected in ancient pottery vessels. Those pots and tankards showing four legs generally mimic some form of animal life. The problem with four legs is that you had better have a level surface to set your vessel down upon or to place your chair on if you don't want the contents sloshing about or yourself rocking in an uncontrolled way.

7. Robert Pirsig, *Zen and the Art of Motorcycle Maintenance* (New York: William Morrow, 1974).

8. Note how some sort of abstract sketch or figure would be extremely useful for describing the chair's principle of operation. Not a glossy picture of a Le Corbusier chaise lounge, but a simple, geometrical figure with a few lines for legs, the outline of the plane figure whose vertices are located at the four points where the feet hit level ground, a special symbol for center of gravity, and finally a line drawn from the center of gravity, through the plane at the base of the feet, directed toward the earth's center. For stability, the intersection of this line with the plane at the feet must lie within the interior of the plane figure.

9. Galilei Galileo, *Discorsi e dimastrazioni matematiche, intorno a due nuoue scienze*, (Leida; Appresso gli Elsevivii, 1638). For a standard translation, see *Dialogues Concerning Two New Sciences*, trans. Henry Crew. Alfonso de Salvio. (New York: Macmillan, 1914), p. 116 (fig. 17).

10. Husserl attributes to Galileo the conception of nature as, in reality, possessing a mathematical structure behind the appearances. The full Husserl quote is "The *Ideenkleid* (the ideational veil) of mathematics and mathematical physics represents and [at the same time] disguises the empirical reality and leads us to take for True Being that which is only a method." Quoted in Robert Cohen and Marx Wartofsky, eds., *Boston Studies in the Philosophy of Science*, vol. 2 (New York: Humanities Press, 1965), p. 286. Or, as put by Aron Gurwitsch in his "Comments on Marcuse" in the same volume (p. 293), "The world is not in reality as it looks."

See also Michael Lynch, "The Externalized Retina: Selection and Mathematization in the Visual Documentation of Objects in the Life Sciences," in Michael Lynch and Steve Woolgar, eds., *Representation in Scientific Practice* (Cambridge, MA: MIT Press, 1990), pp. 153–186, especially pp. 169ff.

11. Even anonymity is not enough to ensure that we have an appropriate response to the question. Take an electric chair. Clearly, this is not particular in the "one-of-a-kind," museum-item sense (although no doubt some wax museum has one on display). How does an electric chair work? What is an electric chair? Our abstract way of answering would have to move beyond Galilean mechanics and refer to voltages and currents, resistive paths through the human body as well as the location of its center of gravity. But, again, we don't quite have it right.

12. Ian Hacking, *Representing and Intervening* (Cambridge: Cambridge University Press, 1983).

13. This line of criticism is forcefully presented in Langdon Winner, *The Whale and the Reactor* (Chicago: University of Chicago Press, 1986).

14. Claude Lévi-Strauss, "On Manipulated Sociological Models," *Bijdvagen Tot de Taal-Land-En Volkenkunde* 116, 1 (1960): 53.

15. This is analogous to criticisms made of philosophical expositions of the scientific method. Although these explanations of meaning and contingency of one relationship with respect to another make sense, there is little evidence that they can be read as a valid account of doing science. See Thomas Kuhn, *The Structure of Scientific Revolutions*, 2d ed. (Chicago: University of Chicago Press, 1970), especially chapters 11 and 13.

16. Of course, if we have a particular artifact in hand, we can validly ask, "Who made *this* chair?" We might be able to label it with a style (Tudor), a century (16th), perhaps even a crafter's name. But when we think of "chair" as generic, abstract, floating above time, we then lose our hold of history, of cultural process, and of situated human activity. For an interesting essay on the innovative process—an essay in tune with the perspective taken here—see Jean Louis Maumoury's summary of the "heroic" and "social" images of innovation process in *La Génèse des innovations* (Paris: Presses Universitaires de France, 1968).

17. Though I speak of "a consumer" in the singular, it is no one consumer acting alone that is the utilitarian's concern. Rather, the term *consumer* is to be taken as a statistical aggregate of all consumers having a similar need for a refrigerator, for example, or a nuclear power plant.

18. Arthur, J. Pulos, *American Design Ethic: A History of Industrial Design to 1940* (Cambridge, MA: MIT Press, 1983), p. 251.

19. The importance and interplay of empirical and social components in knowledge production in science is argued in David Bloor, *Knowledge and Social Imagery* (London: Routledge & Kegan Paul, 1976).

20. Heather Lechtman and Arthur Steinberg, "The History of Technology: An Anthropological Point of View," in G. Bugliarello and D. B. Doner, eds., *The History and Philosophy of Technology* (Urbana: University of Illinois Press, 1979), p. 137.
   The inclusion of "science" as a culturally specific category, rather than as an independent technical constraint, like the physical properties of the material itself, muddies the dialectical waters. Are not physical properties a matter of the culture's science? I would have preferred, for the sake of argument, that they had put science on the other side of the fence, in with the hard realities of energy processes and material properties.
   Note, too, how difficult it is to escape the instrumental way of thinking. To ask "how much. . . ?" invites abstraction and the aggregation of particulars into the generic that can be counted.

21. Donald MacKenzie and Graham Spinardi, "The Shaping of Nuclear Weapon System Technology: U.S. Fleet Ballistic Missile Guidance and Naviga-

tion: I: From Polaris to Poseidon," *Social Studies of Science* 18 (1988): 419–463. Also, more recently, Donald Mackenzie, *Inventing Accuracy: A Historical Sociology of Nuclear Missile Guidance* (Cambridge, MA: MIT Press, 1990).

22. Other recent works focus on the day-to-day work of scientists within and without the laboratory. See Bruno Latour and Steve Woolgar, *Laboratory Life: The Social Construction of Scientific Facts* (London and Beverly Hills: Sage, 1979); H. M. Collins, "The Seven Sexes: A Study in the Sociology of a Phenomenon, Or the Replication of Experiments in Physics," *Sociology* 9 (1975): 205–224, and *Changing Order: Replication and Induction in Scientific Practice* (Beverly Hills: Sage, 1985); K. D. Knorr-Cetina, *The Manufacture of Knowledge: An Essay on the Constructivist and Contextual Nature of Science* (Oxford: Pergamon, 1981); Michael Lynch, *Art and Artefact in Laboratory Science: A Study of Shop Work and Shop Talk in a Research Laboratory* (London: Routledge and Kegan Paul, 1985); A. Pickering, *Constructing Quarks—A Sociological History of Particle Physics* (Chicago: University of Chicago Press, 1984); Trevor J. Pinch, *Confronting Nature: The Sociology of Solar-Neutrino Detection* (Dordrecht: Reidel, 1986); and Sharon Traweek, *Beamtimes and Lifetimes: The World of High Energy Physicists* (Cambridge, MA: Harvard University Press, 1988).

Chapter 2

1. The classic ethnographies include Bronislaw Malinowski, *Argonauts of the Western Pacific* (New York: Dutton, 1961); Margaret Mead, *Coming of Age in Samoa* (New York: Morrow, 1928); and Edward E. Evans-Pritchard, *The Nuer* (Oxford: Oxford University Press, 1940).

2. We might think of dropping the word *subculture* altogether and use instead the language of "roles"; see Erving Goffman, *Strategic Interaction* (Oxford: Basil Blackwell, 1970). But role playing doesn't point in the direction I want to go. It focuses too narrowly on the individual and suggests that the framing of action and exchange, the script and staging, are given. I want to push further and explore how the script is written, how backstage and upstage are defined within the firm. Life there appears as a more spontaneous activity than Goffman would have it, as much a shifting of scenery and ad-lib activity as a codified performance.

3. Two discursive bibliographies on the topic of organizational culture from a management perspective are G. A. Fine, "Negotiated Orders and Organizational Culture," *Annual Review of Sociology* 10 (1984): 239–262; and W. G. Ouchi and A. L. Wilkins, "Organizational Culture," *Annual Review of Sociology* 11 (1985): 457–483.

4. Studies of the practices and productions of scientists at work are a relatively recent phenomenon, a matter of the past two decades. What dates and distinguishes the prior analyses of sociologists from those of today is that the former, in their sweeping conjectures about the dependence of science on particular

types of social structures, are unable to demonstrate (or do not see the need to show) any direct causal connection between their descriptions of social order or prevailing ideologies and the productions of the scientists. While their ideas may ring true, their stories imply that science could not be different from what so neatly resonates with their thesis. In short, they take the process of doing science at the local level at face value and as inconsequential in their analyses. This stance conveniently maintains the professional disjunction of the sociologist from the scientist toiling away in the laboratory. It is not surprising that the scientist then sees nothing of practical import in the sociologist's message. Both sociologist and scientist can go about their business uncontaminated by the other. Each respects the other because it doesn't matter what the other claims as truth. Prior to Kuhn, Lakatos, and Feyerabend, historians of science were ignored in much the same way.

Today's sociologists, on the other hand, in their micro studies of scientific practice in "real time," actively question the interplay of craft technique, modes of thought, values, and interests in the day-to-day elaboration of theory, construction of evidence, and maintenance of the enterprise as a whole. Wholesale dismissal of the implications of these more contemporary studies requires more effort. Early, provocative works in this vein include Bruno Latour and Steve Woolgar, *Laboratory Life* (Beverly Hills, CA: Sage, 1979); and K. D. Knorr-Cetina, *The Manufacture of Knowledge: An Essay on the Constructivist and Contextural Nature of Science* (Oxford: Pergamon, 1981).

The social study of technology in this mode is yet a more recent development. See the book that gave birth to the Inside Technology series: Wiebe E. Bijker, Thomas P. Hughes, and Trevor Pinch, eds., *The Social Construction of Technological Systems* (Cambridge, MA: MIT Press, 1987).

5. Malinowski's observation of the Trobriand Islanders' engineering of a canoe formed the basis for his classic treatise. For a more recent exploration of the significance of material ingredients in the life of a scientific culture, see Sharon Traweek, *Beamtimes and Lifetimes: The World of High Energy Physicists* (Cambridge, MA: Harvard University Press, 1988).

6. Edgar H. Schein, "The Role of the Founder in Creating Organizational Culture," *Organizational Dynamics* (Summer 1983): 13–28.

7. Our perspective aligns with that of R. Rosaldo: "In contrast with the classic view, which posits culture as a self-contained mode of coherent patterns, culture can actually be conceived as a more porous array of intersections where distinct processes criss-cross from within and beyond its borders" (*Culture and Truth: The Remaking of Social Analysis* [Boston: Beacon Press, 1989], p. 20).

8. There are multitudes of other techniques, agents, and artifacts essential to commuting. Consider all the apparatus of a supporting infrastructure that makes commuting possible (or impossible): parking garages, tollbooths, traffic police, service stations, gas delivery systems, roadways and their maintenance crews, parts suppliers, even junkyards. All of this and more is part of the technology of commuting by automobile.

9. Here begins what anthropologists call an "arrival story," a common device of ethnographic writing that positions the reader relative to both the field-worker and the fieldwork. See M. L. Pratt, "Fieldwork in Common Places," in J. Clifford and G. E. Marcus, eds., *Writing Culture* (Berkeley: University of California Press), pp. 27–50.

10. The deficiencies of an ahistorical perspective, a stance usually assumed by ethnogrophers, are neatly summarized in chapter 3 of J. Van Maanen, *Tales of the Field* (Chicago: University of Chicago Press, 1988), p. 72n17. A fuller discussion of attempts to move toward "historizing the ethnographic present" is found in G. E. Marcus and M. M. J. Fischer, *Anthropology as Cultural Critique* (Chicago: University of Chicago Press, 1986), to which Van Maanen refers.

11. Van Maanen, in *Tales of the Field*, notes how the field-worker's own standards for what is relevant and a host of other factors condition what he or she sees as data, and he uses Paul Ricoeur's term "textualization" to describe this process "by which unwritten behavior, beliefs, values, rituals, oral traditions, and so forth, become fixed, atomized, and classified as data of certain sort." See Paul Ricoeur, "The Model of the Text," *New Literary History* 5 (1973): 91–120.

12. You might "test" the account through the coherence of the entire text. But a novel is coherent, and if there is any truth to the saying about truth being stranger than fiction, then coherence in an ethnography may be a sign that it is fictive. As to the judgment of participants themselves: "Of course there are problems here. For example, there may be more than one correct account, and group members may disagree in any case on which account or accounts are correct" (M. H. Agar, *The Professional Stranger* [New York: Academic Press, 1980], pp. 78–79). The multiplicity of correct accounts is a probability we address in a later chapter.

13. "Ethnomehodologists" take this microscopic view of process. See H. Garfinkel, *Studies in Ethnomethodology* (Englewood Cliffs, NJ: Prentice-Hall, 1967); and W. W. Sharrock and B. Anderson, *The Ethnomethodologists* (New York: Tavistock, 1986). For a specific example of the kind of detailed analysis of scientific activity that is possible with an audio record see H. Garfinkel, M. Lynch, and E. Livingston, "The Work of a Discovering Science Construed with Materials from the Optically Discovered Pulsar." *Philosophy of the Social Sciences* 11, 2 (1981): 131–158.

14. Clifford Geertz describes the possible meanings of what's in a wink. One learns that even being there to observe the event does not ensure a correct reading. See his *Thick Description: Toward an Interpretive Theory of Culture* (New York: Basic Books, 1973).

15. Max Weber, *Essais sur la Théorie de la Science* (Paris: Libraire Plon, [1922] 1965) (the original German edition is included in *Gesammelte Aufsätze zur Wissen*). See also Gunnar Myrdal, *Objectivity in Social Research* (New York: Pantheon, 1969), pp. ix–xvi.

16. As Robert Pirsig puts it, there is a knife moving here, "a methodological preliminary to revealing the underlying form of the workings of technology" (*Zen and the Art of Motorcycle Maintenance* [New York: William Morrow, 1984], p. 79).

17. At least one philosopher has questioned this way of thinking and conceptualizing the object as *chemical element*. But without this positing of structure, I believe it is impossible, through empirical test alone, to distinguish element from compound in a given set of materials using the "no further reducibility" criterion, even granted all possible reactions. See F. A. Paneth, "The Epistemological Status of the Chemical Concept of Element, II," *British Journal for the Philosophy of Science* 13, 50 (August 1962): 144–160.

Chapter 3

1. This is in some ways analogous to the recognition of theory-laden facts in the doing of science. For theory construction in conjunction with fact definition, see N. R. Hanson, *Patterns of Discovery: An Inquiry into the Conceptual Foundations of Science* (Cambridge: Cambridge University Press, 1965). An engineering example of "theory-fact" development is well illustrated in Walter G. Vincenti, "The Air-Propeller Tests of W. F. Durand and E. P. Lesley: A Case Study in Technological Methodology," *Technology and Culture* (1979): 712–751.

2. For a similar construction, see Nelson Goodman, *Ways of Worldmaking* (Indianapolis: Hackett, 1978). There is also a resonance here with the concept of *boundary object* as described in Susan Leigh Star, "The Structure of Ill-Structured Solutions: Boundary Objects and Heterogeneous Distributed Problem Solving," Eighth AAAI Conference on Distributed Artificial Intelligence, May 23–25, 1988.

3. Christopher Alexander, *Notes on the Synthesis of Form* (Cambridge, MA: Harvard University Press, 1964). We read of design as a process of "proposing" and "disposing" of alternatives, discarding those that won't fit. While the process is never as clearly segmented into these two stages, the proposing and disposing can be read as the story-making and story-changing process described here.

4. The impossibility of completely verifying a design and the relation of this to the challenge of doing design is akin to Karl Popper's argument for the impossibility of completely verifying the constructions of scientists and how that makes what scientists do scientific. See *The Logic of Scientific Discovery* (New York: Basic Books, 1959).

Chapter 4

1. It is this characteristic that stands behind historians' faith in the decisiveness of the "internal logic" argument for explaining design decisions—that is, one can deduce from the internal logic of the structure and function of the artifact

how it was designed as well as why it takes the form it does. To me, the phrase *internal logic* attributes too much to the artifact, as if it possessed and expressed this logic alone. I prefer the phrase *underlying form,* which implies more of a distance between the concepts and principles of science and the working machine. Indeed, my argument is that different participants rely upon different formalisms in designing one and the same artifact. The object embodies multiple underlying forms. The term *multiple internal logics* strikes me as contradiction.

2. This point is argued more fully in L. L. Bucciarelli and D. Schön, "Dialogue Concerning at Least Two Design Worlds," *Proceedings of the NSF Workshop in Design Theory and Methodology,* Oakland, California, 1987.

3. How one might manage the functioning of an artifact without going "inside" to see what explains its behavior is shown in Bruno Latour, *Science in Action* (Cambridge, MA: Harvard University Press, 1987).

4. Along with the hierarchy of knowledge comes a social and professional status structure within the firm. The scientist from Product Development who is responsible for knowing how x-rays behave, how electron-hole pairs move through a photovoltaic cell, or how an E&M device works demands higher esteem (and pay) than the drafter who designs the shielding for the x-ray inspection machinery on a CAD computer or the technician who wires the photovoltaic lighting system before shipment, or the technician who tests the E&M device in the lab.

5. This again is a process of abstraction of sorts—one that leaves aside all that does not fit within this binary juxtaposition. See Herbert Simon, *Sciences of the Artificial,* 2d ed. (Cambridge, MA: MIT Press, 1981), for the notion of *satisficing* designs.

6. Indeed, the "correct" form of the language of science as it appears in the texts is itself a fabrication. The language of science is a "living" language in spite of the stated norm. The meanings of words like *force* are not static, fixed, and bounded by a dictionary or philosopher's scalpel, but variable and suggestive. Through the richness of language, the principles of science are not only made available for teaching the young, but also enable the construction of new science. For example, the story that heat flows from a hot body to a cold body like a material substance is a metaphor that is not only useful in the classroom, but also inspired Sadi Carnot in the first half of the nineteenth century to investigate with profit—intellectual, scientific profit as awarded by future generations—the limits on the amount of work one can get out of a "heat engine" such as a steam, gasoline, or diesel engine. But in further studies of thermodynamics, the student learns that heat is the rapid oscillatory motion of very small entities, molecules, and if they open the Pandora's box of the history of science, they find that Carnot's fluid-flow metaphor exacerbated the confusion of several decades of scientists. Just what *molecule* may mean is another story.

Newton himself is responsible for this laxness, or should we say that we owe him homage because he was one of the first to see how to exorcise the unneeded (and unwanted) spirits from Natural Philosophy while at the same time leaving the door ajar for future generations to traffic in demons of their own making. Newton would appreciate Don's story about the air knife: The air flowing in a jet onto the paper is deflected, its momentum is changed, and a force, as a result, is exerted on the paper by the air jet. Don speaks of the momentum change as the cause of the force. But the devil's advocate might claim: If Don takes away the paper, there is no force, so why isn't the paper a coequal "cause" of the force? Newton, and here is where his eminence shows, allowed this to be so; in fact, the principle of "to every action there is an equal and opposite reaction" is enshrined as one of his three fundamental axioms. But what "sense" does it make to speak of cause and effect if, to every action, there is immediate contemporaneous equal and opposite reaction? What is cause, what is effect, other than a fabrication, a reflection of our human inclination to impute order, direction, and motive to the inanimate world? Indeed, action/reaction shows the same inclination. The scientific revolution owes as much to this kind of (linguistic) rhetorical innovation as it does to observation, mathematics, etc.

7. Aristotle, "Poetics," in *The Complete Works of Aristotle,* vol. 2 (Princeton, NJ: Princeton University Press, 1989).

8. Latour emphasizes the importance, indeed priority, of two-dimensional images that simplify confusing three-dimensional objects in the work of scientists. See Bruno Latour, "Drawing Things Together," in Michael Lynch and Steve Woolgar, eds., *Representation in Scientific Practice* (Cambridge, MA: MIT Press, 1990), p. 39. The circumstances here are different. In design, the object doesn't exist as a three-dimensional artifact, although bits and pieces lie about.

9. Others have stressed the importance of visual thinking in design and engineering. See Eugene F. Ferguson, *Engineering and the Mind's Eye* (Cambridge, MA: MIT Press, 1992). I chose figures 2–5 to emphasize the importance of visual thinking about matters other than spatial relationships.

10. Bruno Latour argues for the importance of a "practical set of skills to produce images, and to read and write about them" (I would add "speak about them") in the work practices of scientists: "We need . . . to look at the way in which someone convinces someone else to take up a statement, to pass it along, to make it more of a fact, and to recognize the first author's ownership and originality" ("Drawing Things Together," pp. 22, 24). Kathryn Henderson has argued for the critical importance of the sketch (as boundary object) in the design process and suggests that attempts to improve design process through the introduction of computer aids are destined for failure if the social functions of the sketch are not seen; see "Flexible Sketches and Inflexible Data Bases: Visual Communication, Conscription Devices, and Boundary Objects in Design Engineering," *Science, Technology, & Human Values* 16, 4 (Autumn 1991): 448–473.

11. The transient nature of images as data objects in science and their negotiation is described in K. Amann and K. Knorr Cetina, "The Fixation of (Visual) Evidence," in Lynch and Woolgar, eds., *Representation in Scientific Practice*, p. 92.

12. See Aaron Cicourel, "Ethnomethodology," in *Cognitive Sociology* (New York: Free Press, 1974), for a provocative discussion of the importance of context in the fixing of meaning. See also his *Method and Measurement in Sociology* (New York: Free Press, 1964).

13. Learning to "read" a drawing can be a lengthy process. I still cannot "see" the face in the black-and-white image presented in H. Collins, "An Empirical Relativist Programme in the Sociology of Scientific Knowledge," in K. Knorr-Cetina and M. Mulkay, eds., *Science Observed: Perspectives on the Social Study of Science* (London: Sage, 1983), 85–114. Once you do see without prompting, once you have learned to sketch in the language of the object world, it is very difficult to go back and *not* see the drawing as it was intended.

14. James L. Meriam, *Engineering Mechanics: Statics and Dynamics*, vol. 2, *Dynamics* (New York: Wiley, [1966] 1978), p. 66.

15. The relation between textbook knowledge and the practical, hands-on know-how of the tinkerer and builder is an important and complex question. Walter Vincenti has written several articles addressing the status of engineering knowledge in different contexts; see, for example, "Control-Volume Analysis: A Difference in Thinking between Engineering and Physics," *Technology and Culture* (1982): 145–174.

16. This is where the problem lies: the apparent contradiction of the physical figure, which clearly won't allow point A to move with constant velocity forever, and the statement that the particle is indeed moving with constant velocity. Thus, the felt need to limit the time to an instant. Contrast this with Galileo's argument about how an object thrown upward must pass through all degrees of speed. At the top, the velocity is zero for an instant (but not constant). How can it go anywhere if its velocity is zero? The text thus indicates ambivalence, still a problem with the concept of limit.

Note, too, how the hint now becomes clear and appropriate. We have reduced the problem statement down to the fundamental level of the hint. The hint itself need not be "deconstructed" in the least.

17. As one includes more irrelevant information, one is specializing the situation, making the problem less general; that is, the mechanism as shown might be used in a wide variety of ways. So one could argue that this is why you don't add more irrelevant information. It would make the problem too particular, and the student wouldn't recognize the general power of analysis. But if you carry that argument to its logical conclusion, *no* irrelevant information should be included, and we once again arrive at the same conclusion.

18. John Dixon, *Design Engineering* (New York: McGraw-Hill, 1966). For the more elaborate representation, see Morris Asimow, *Introduction to Design* (Engle-

wood Cliffs, NJ: Prentice-Hall, 1962), where the author uses fully seven times more "stages" to describe the sequence of tasks that he sees performed in taking a design from inspiration to production.

19. Frederick Taylor, *Principles of Scientific Management* (New York: Norton, 1967).

20. One can imagine a computer-generated version of this block diagram with iconic hands processing and passing along the necessary forms, suggesting, but not revealing, human agents. Roland Barthes, "The Plates of the *Encyclopedia,*" in *New Critical Essays* (New York: Hill & Wang, 1980).

21. L. Bucciarelli, "Is Idiot Proof Safe Enough?" *International Journal of Applied Psychology* 2, 4 (Fall 1985).

22. See Julian E. Orr, "Talking about machines: Social Aspects of Expertise," Xerox Palo Alto Research Center, May 1987.

Chapter 5

1. The dynamics of infrastructural change has serious implications for the lives of skilled machinists. The possibility of alternatives in design here is argued in David F. Noble, *Forces of Production: A Social History of Machine Tool Automation* (New York: Knopf, 1984).

2. Constraints can also be argued as a resource. Without constraints, design is impossible. See L. L. Bucciarelli and D. Schön, "Dialogue Concerning at Least Two Design Worlds," *Proceedings of the NSF Workshop in Design Theory and Methodology,* Oakland, California, 1987.

3. Where is the interface between rooftop and photovoltaic module? Negotiable!

4. Henry Petroski, *To Engineer Is Human: The Role of Failure in Successful Design* (New York: St. Martin's Press, 1985).

5. Putting together a project team dedicated to a particular design task, all the while maintaining the hierarchy of the firm, gives rise to another level of complexity. Here a matrix representation finds use: Rows are different design tasks, columns agents in the firm's organization. The difficulties of pulling off smoothly this dual allegiance are well known.

6. M. Mauss, *The Gift: Forms and Functions of Exchange in Archaic Societies* (Glencoe, IL: Free Press, 1954). Also, in Randall Collins, *Sociology since Midcentury* (New York: Academic Press, 1981), we find a proposed model or theory of chained interactions to explain, not just describe, the microsocial behavior of individuals and small groups—an explanation that involves the "emotional" as well as material interests of members of a group. This is evident when one looks within the firm at the ecology of the design team. Here the "infrastructure" has

an immediate local dimension and connections of exchange with all the resources of the firm.

Chapter 6

1. Stuart Pugh, "Concept selection—A method that works," *Proceedings of the International Conference on Engineering Design*, Rome, 1981, pp. 447–506. The method is similar to Christopher Alexander's vision of designing as a process of eliminating those features that don't fit. See Christopher Alexander, *Notes on the Synthesis of Form* (Cambridge, MA: Harvard University Press, 1964).

2. Another possibility is that a set of specifications prompt no conflict because participants are not ready to deal seriously with the criteria, or they are so vaguely stated that conflicting interpretations do not come to the surface. On the other hand, if all participants appear to understand and value a detailed presentation in the same way, this implies that all participants, from diverse object worlds, have a common reference in mind in doing their reading. This, in turn, suggests that there is very little freedom left in the generation of a design. It suggests a fait accompli, a superficial revision of last year's model or the like. This is not designing.

3. Engineers display tendencies to want to fill in all the cells of the matrix, as if only then have they captured the full complexity of the relationships among the entries. But note that the fabrication of this matrix still constrains and limits the kinds of relationships that can be represented. The lines between rows and between columns tend to falsely isolate one row or column entry from another. There are "hard" taxonomic techniques for clustering and determining levels of independence among the entries in a matrix such as this. Their use in developing the Pugh matrix might lead to some interesting insights—as much into the process of classifying (what's really going on here) as they might prove useful to the method's practitioners. For taxonomic method, see Robert Q. Sokol and Peter H. A. Sneath, *Numerical Taxonomy* (San Francisco, CA: W. H. Freeman, 1973).

4. For a critique of "interests" used to explain the actions of scientists, see Steve Woolgar, "Interests and Explanation in the Social Study of Science," *Social Study of Science* 11, 3 (1981): 365–394. A response follows in the next number: S. B. Barnes, "On the hows and whys of cultural change," *Social Study of Science* 11, 4 (1981): 381–398.

My interests are more directly tied to a participant's responsibilities in design and intimately wrapped up with their work, both social and cognitive, within object worlds. Interests are part of the territory, so to speak. I find Jonathan Turner's description of the diverse ways scholars treat norms to be sympathetic: "I see norms as a process revolving around validating, accounting, and role taking. As people negotiate over what is proper, authentic and efficient (validating), as they negotiate over the proper interpretive procedures or 'ethno' methods for creating a sense of common reality (accounting), and as they try to put themselves in others' places and assume their perspective (roletaking), they *do* develop implicit and provisionally binding agreements about how they are to interact and adjust their conduct to each other" ("Analytical Theorizing," in Anthony Giddens and Jonathan Turner, eds., *Social Theory Today* [Stanford, CA: Stanford University Press, 1987], p. 181).

5. I speak of "a customer" in the singular. This is shortsighted. Customers also, with or without the aid of the media, purchase, use, and discard a firm's products within contexts such that the artifact's worth and performance are judged by socially constructed expectations, experiences, and grievances as much as by the instrumental measures of an edition of *Consumers Report*—so limited in its insight—or the manufacturer's documentation, equally deficient.

6. The argument here resonates with Imre Lakatos's classical argument concerning the impossibility of verifying Euclid's theorem. The way engineering problem solvers manipulate and maintain a problem definition in their work of fabricating solutions is akin to the strategies mathematicians have used to maintain Euclid's theorem. See Imre Lakatos, *Proofs and Refutations: The Logic of Mathematical Discovery* (Cambridge: Cambridge University Press, 1976).

7. For the degrading effect of mismatch in the performance characteristics of individual cells upon the quality of the photovoltaic module as a whole, as measured by its efficiency in converting the sun's radiant energy into electricity, see Louis L. Bucciarelli, "Power Loss in Photovoltaic Arrays due to Mismatch in Cell Characteristics," *Solar Energy* 23 (1979): 277–288.

8. Attempts to break the design task into independent subtasks that might be addressed in parallel, or to which one might assign relative priorities, are not without merit. See Nam Suh, *The Principles of Design* (New York: Oxford University Press, 1990), for the prescription of a method. But this example makes it clear that it will be extremely difficult to draw boundaries around wholly uncoupled tasks. The perspective adopted here is that any attempt at drawing boundaries around subsystems or features involves making presumptions about what the object of design is and ought to be and, as such, is in itself a design act.

9. Here we seek, following Wittgenstein, "to understand the relational character of signification in the context of social practices" (Anthony Giddens, "Structuralism, Post-Structuralism and the Production of Culture," in Giddens and Turner, eds., *Social Theory Today*, p. 204). For "key symbol," see Sherry B. Ortner, "On Key Symbols," *American Anthropologist* 75 (1973): 1338–1346.

10. Nominalization is often the means used in this creative process: The things that get named have function; they are the active agents in object-world story land. Making a label by transmuting a verb of action into a noun solidifies the function as an object.

Acronyms work in the same way. They are favorite fabrications of engineers. The acronym reifies in the same way by establishing an identity where only function exists. A string of four or five uppercase letters carries considerable authority once defined. The search for the most clever expression can become a design task in itself.

An example of how a special symbol can sustain the perception of an event or function in-the-making is found in Steve Woolgar's discussion of the identification of the "glass transition temperature, T_g," in the conduct of a laboratory test. See "Time and Documents in Researcher Interaction: Some Ways of Making Out What Is Happening in Experimental Science," in Michael Lynch and Steve Woolgar, eds., *Representation in Scientific Practice* (Cambridge, MA: MIT Press, 1990), pp. 123–152, especially p. 141.

11. Ruth P. Mack, *Planning on Uncertainty* (New York: Wiley Interscience, 1971), p. 6.

Chapter 7

1. The work of scientists themselves traditionally has been depicted as the rational evolution of ever more comprehensive theory buttressed or denied by experiment. While alternative paths and paradigms are admitted, the decision-making process classically has been historicized as uncontaminated by social constructs. But if you open up the boxes built by popular historical and philosophical commentators, as some social scientists have done over the past few decades, you begin to see a process much like designing within the firm, a social process of negotiated order. The fault is in the scope of the model, their breadth of view, not in the rigor of the reasoning or the soundness of historical data.

2. But you have to go there and watch the process to capture this essential feature of designing. It is not recoverable from the artifact, nor is it there in the documentation and textual remnants of the design process. These, in their object-world voice, speak only of the empty shell of designing. Oral histories are a potentially sufficient source; these will also take the form of object-world representations of process. Multiple oral histories, if my thesis is correct, ought to be conflicting, reflecting the different interests and responsibilities of participants. The synthesis of a coherent story about design process from these becomes then a process of negotiation.

3. From an analytical perspective, the event prompts reflection on two fundamental questions in social studies: How is social order sustained in a less than rigidly authoritarian system? How do we relate the micro study, the ethnography of design within the firm, to the macro study of historical events, political currents, and processes ever swirling around?

We have seen that the social order within the design team is a negotiated order. So, too, organizational theorists have argued that the essence of the operation of the firm as a whole is negotiation. See A. Strauss, *Negotiations* (San Francisco: Jossey-Bass, 1979); and D. Maines and J. Charlton, "The Negotiated Order Approach to the Analysis of Social Organization," in H. Fabermen and R. Perinbanayagam, eds., *Studies in Symbolic Interactions,* Suppl. 1, *Foundations of*

Interpretive Sociology, 1985, pp. 271–308. Apparently, then, we can have multiple levels of negotiated order. But how do we account for the disjunction between levels? What happened at Solaray was certainly not the outcome of a negotiation among all interested parties. How can we explain and contain this event, and others like it, without acknowledging the critical significance of something more than negotiation as the basis for designing as a social process? A more authoritarian and dictatorial process is apparent.

With regard to the second fundamental question, I have been able to accommodate the apparent chaos made evident in the design process within Solaray by the conceptual apparatus of "object worlds" and by positing design as negotiation within the *microworld* of the firm. But now my microworld is shaken from its very foundation! To encompass this event I might then, instead of attempting to broaden the scope of my study as suggested above—defining a second subculture, positing connections between corporate designing and photovoltaic technology designing, and researching and then constructing an account that stands apart from history (another object world, timeless story)—acknowledge the event as unique and depart from my synchronic analysis, step out of the book with its vignettes woven together with little attention to strict chronology, and make a true history of Solaray that is solidly grounded in the macroworld of the 1980s. For the problem of connecting micro and macro renderings, see K. D. Knorr-Cetina and A. V. Cicourel, eds., *Advances in Social Theory and Methodology, Toward an Integration of Micro- and Macro-Sociologies* (London: Routledge & Kegan Paul, 1981).

4. The seamless web of social and instrumental factors and agents in technological development is defined and illustrated in Thomas P. Hughes, *Networks of Power: Electrification in Western Society, 1880–1930* (Baltimore: Johns Hopkins University Press, 1983). For the heterogeneous nature of engineering processes see John Law, "Technology and Heterogeneous Engineering: The Case of Portuguese Expansion," in Wiebe E. Bijker, Thomas P. Hughes, and Trevor Pinch, eds., *The Social Construction of Technological Systems* (Cambridge, MA: MIT Press, 1987).

5. I am reminded of those large atlases of human anatomy that show the various subsystems of the human body in rich and colorful detail. Indeed, if we think of the human body as an artifact, it is easy to identify the different object worlds—circulation of the blood, the nervous system, the lungs, digestive subsystem, the skeleton, muscles—that would have to be attended to in its design.

6. Think of Robert Pirsig's description, in *Zen and the Art of Motorcycle Maintenance* (New York: William Morrow, 1974), of how his motorcycle develops its "personality." At one time it was just like all the others, on the production line. But in his "care," it grows, adapts, and responds to his every move; the clutch plate wears to the touch of his control, the battery discharges and sets its internal state in response to his strategy for warm-up, the brakes too are conditioned to the timing and pressure of his actions. And in repairing his machine, the torque he applies to a bolt stresses the steel just slightly differently from what another rider might do. The user appropriates the artifact, makes it

his or her own, tacitly shapes it according to his or her values, interests, and expertise. Here is object-world activity past the design stage.

7. Much of the argument over questions concerning the distinction between science and technology, whether engineering is applied science, and the like, could be avoided if a similar displacement were made and attention were directed to the social actors and agents who do science and engineering in their organized and institutional settings. No doubt multiplicities of subcultures, contexts, and groupings might be discovered or constructed, but this apparent complexity leads to a much richer analysis than attempts to reify science or technology at some more etherial, timeless, socially, historically, and politically disjunct level.

8. In the light of this complexity, some would claim that economic measures are the best to use, arguing that they can capture almost all criteria; that is, everything and everyone has their price, all else being equal. The intrusion of "externalities" into analyses of this sort reveals the difficulty of encompassing the intersections of interests in anything other than the most global of contexts.

9. For a proposed set of criteria to support democratic technique, see R. Sclove, "Technological Politics as if Democracy Really Mattered: Choices Confronting Progressives," in M. Shuman and J. Sweig, eds., *Technology for the Common Good* (Washington, DC: Institute for Policy Studies, 1993).

Index

Printed in the United States
By Bookmasters